● ●

CARBON-FREE AND NUCLEAR-FREE

A Roadmap for U.S. Energy Policy

Arjun Makhijani, Ph.D.

A Joint Project
of the
Nuclear Policy Research Institute
and the
Institute for Energy and Environmental Research
August 2007

Foreword by
S. David Freeman
President, Los Angeles Board of Harbor Commissioners
and former chairman of the Tennessee Valley Authority
and
Afterword by
Dr. Helen Caldicott
Founding President, Nuclear Policy Research Institute

RDR
Books

Published jointly by
IEER Press
Takoma Park, Maryland
and
RDR Books
Muskegon, Michigan

IEER Press

Carbon-Free and Nuclear-Free
A Roadmap for U.S. Energy Policy

published jointly by

RDR Books	IEER Press
1487 Glen Avenue	6935 Laurel Avenue, Suite 201
Muskegon, MI 49441	Takoma Park, MD 20912
Phone: (510) 595-0595	Phone: (301) 270-5500
Fax: (510) 228-0300	Fax: (301) 270-3029
E-mail: read@rdrbooks.com	E-mail: ieer@ieer.org
www.rdrbooks.com	www.ieer.org

ISBN: 978-1-57143-173-8

Library of Congress Control Number: 2007936263

Cover painting: Pushpa Mehta
Cover design and production: Richard Harris
Interior design: Elizabeth Thurlow-Shields and Ken Shields

Distributed in the United Kingdom and Europe by
Roundhouse Publishing Ltd.
Millstone, Limers Lane,
Northam, North Devon EX39 2RG, United Kingdom

Distributed in Canada by
Scholarly Book Services
127 Portland St. 3rd Floor
Toronto, ON M5V 2N4

Printed in the United States of America

To

Helen Caldicott and S. David Freeman

who both inspired this book

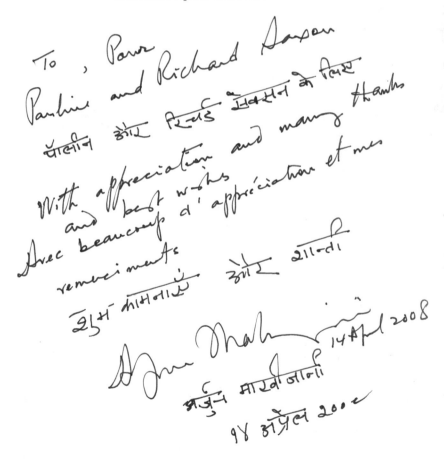

CONTENTS

FOREWORD

by S. David Freeman

In the summer of 1972, Arjun Makhijani walked into my office at 1776 Massachusetts Avenue in Washington, D.C. for an interview. He had a head full of hair and numbers. At the time, I was the Director of the Energy Policy Project of the Ford Foundation. Working in the White House in the 1960s, I felt that U.S. energy policy was seriously adrift and that we would soon run into trouble if oil imports kept rising inexorably. I wanted to lead an effort to change U.S. policy to one that would give us economic growth with much lower energy growth or even zero energy growth. It would free our foreign policy and literally allow us to breathe freer in our cities, which were choking with pollution. Zero energy growth with positive economic growth was considered economic heresy then; the experts believed that economic growth and energy use growth inevitably went hand-in-hand. But some of us saw the crisis coming and the Ford Foundation agreed to set up an internal project to see what could be done. I had the vision for the direction that the country should take. In Arjun, I had found the man with the numbers savvy to help me figure out the efficiency angle.

As a doctoral student at the University of California at Berkeley, he had already done preliminary estimates of the energy efficiency potential of the U.S. economy, two years before the Arab oil embargo. He was the principal author of a seminal 1971 study on energy efficiency with a typically vague and academic title: *An Assessment of Energy and Materials Utilization in the U.S.A.* Arjun's work on energy efficiency soon became the technical core of the demand-side of the "Technical Fix" and "Zero Energy Growth" scenarios that we had set out to construct.

When the energy crisis broke over the United States like a political and economic tsunami in October 1973, our project was the only independent game in town. The country needed answers and we had been asking the right questions. Though much remained to be done, the numbers were ready; we published them in a preliminary report, *Exploring Energy Choices*, in January 1974. That work, and our final report, *A Time to Choose: America's Energy Future*, became the foundation of President Carter's energy policy. I have recounted that story in my own book, *Winning Our Energy Independence*, published by Gibbs-Smith on October 1, 2007.

When President Carter appointed me to the Board of Directors of the Tennessee Valley Authority, and then promoted me to be the Chairman, the country was in the midst of a profound change in its energy consumption patterns. Economic growth had resumed, but energy growth had not. The Zero Energy Growth scenario that the then-President of Mobil Oil Company, William Tavoulareas, had been so critical about (he was on our Board of Advisors) was actually being realized in practice. But TVA had its head in the sand; it was building 14 nuclear reactors at once, as if 1973 had been just another normal year. It was business-as-usual in the worst way.

I wanted someone to advise me on how to put a thorough energy efficiency program into place at TVA. Arjun came to TVA to work with me as a consultant in 1978. Typically, he took a look at the big picture of TVA's supply and demand first. He wrote a report whose gist was that unless TVA cancelled at least eight reactors (he actually named the ones), an energy efficiency program would be counterproductive. It would reduce demand growth when it was already slowing. At the same time the reactors would greatly increase TVA's capacity to generate electricity that would likely have no market. It was a recipe for trouble. I had a long, tough road ahead of me to put TVA's house in order, but by 1982 I did manage to get all eight of them cancelled; I also put in place what was then the country's largest energy efficiency program. Once more, Arjun's analysis was right on target.

With *Carbon-Free and Nuclear-Free*, he has done it again. But this time he had to be goaded into doing the study. Last year, I gave a talk at an energy conference sponsored by Helen Caldicott's Nuclear Policy Research Institute. I said that the United States should jettison both coal and nuclear power. The future lay with solar energy. We should begin the transformation now and finish it as soon as possible. Helen was in enthusiastic agreement. But Arjun came up to me afterwards and said: "You are proposing a course that is so costly that it would drive every industry we have to China." I told him to stop being a naysayer and analyze how we could move from our polluting oil addiction to renewable energy. He didn't believe it could be done, but he agreed to take a preliminary look out of respect for Helen and me. To his surprise, he found there was a technological revolution going on that he had missed, because he focused for so many years on the environmental and health problems caused by nuclear weapons production and testing.

Sharing our concerns about climate change, the risks of nuclear power, and the problems of oil import dependence, he agreed to take up the challenge of examining the feasibility of a renewable energy economy. Helen agreed to raise the money. His very diverse Advisory Board, of which I am a member, critically reviewed the outline of this book and its first draft. He has carefully taken our suggestions into account. He interviewed leaders of established and emerging industries. He reviewed an enormous amount of recent technical literature on energy that seems to have attracted little notice in Washington, D.C. *Carbon-Free and Nuclear-Free* is the result.

This Roadmap could liberate us from an energy policy that is trashing our climate and our mountain tops, that is polluting our land, sea, and air, that is trying to resurrect dangerous nuclear power, and that has America so dependent on imported oil that our foreign policy is the prisoner of oil. It shines a light on the path to a renewable energy economy. It will not be easy to get there, but it can be done. Arjun's head has less hair (he says he has "grown old and bald doing environmental work for thirty seven years") but it is still full of reliable numbers.

My advice in these turbulent energy times is: when Arjun talks numbers, policymakers should listen. He has a stellar technical track record. It is time again to choose. Last time, we achieved zero energy growth with positive economic growth when few thought it was even within the realm of possibility. I have no doubt that, with determination and guts, we can achieve a renewable energy economy. Arjun has laid out a thoughtful and practical approach to get us there.

S. David Freeman
President, Los Angeles Board of Harbor Commissioners
August 2007

ACKNOWLEDGEMENTS

I would like to thank the Nuclear Policy Research Institute (NPRI) for having sponsored the project that resulted in this book. Dr. Helen Caldicott was the star who raised the funds, provided critical comments and suggestions, and, most of all, had the vision that this study should be done because it is urgently needed. I am deeply grateful to her.

The project was conceived at NPRI's 2006 energy conference, where S. David Freeman, who led the Energy Policy Project of the Ford Foundation in the early 1970s, was among the speakers. He presented a vision for a solar energy future that, along with Helen's vision for this project, inspired this book.

This project to create a Roadmap for a zero-CO_2 emissions, non-nuclear energy future for the United States was immensely helped by the insights and comments of an Advisory Board. Only the names of those who were able to participate are included:

1. Helen Caldicott, President, Nuclear Policy Research Institute, which is sponsoring this project.
2. Paul Epstein, Associate Director, Center for Health and the Global Environment, Harvard Medical School
3. S. David Freeman, President, Los Angeles Board of Harbor Commissioners and former chairman of the Tennessee Valley Authority
4. Dawn Rittenhouse, Director, Sustainable Development, DuPont
5. Jenice View, Education and Training Director, Just Transition Alliance
6. Hisham Zerriffi, Ivan Head South/North Chair at the University of British Columbia

They provided extensive comments on an outline of this report and on a draft of the full report. They attended an all-day meeting on March 15, 2007, during which they provided criticisms and suggestions for the final product and ideas for outreach. I also corresponded with Hisham and Dawn on several technical topics. John Carberry of DuPont provided very helpful comments; he and Dawn did an interview with me, which is reproduced as Appendix B of this book, which provided insights on industrial energy efficiency, among other things. I am very grateful to them.

I wish to note here that DuPont's position on nuclear power is to keep that option open, in contrast to the recommendation in this book that it be phased out. Their position on carbon caps is also different. Please refer to http://www.us-cap.org/USCAPCallForAction.pdf for details.

While the comments have helped to improve this work greatly, as the author of this work, I alone am responsible for any remaining errors or omissions and for the analysis, findings, recommendations in it.

Mark Selden, editor of *Japan Focus* (www.japanfocus.org) , interviewed me after the summary of this book was published on July 29, 2007. His questions were very sharp and practical. The interview sets this book in an international context and is republished here as Appendix C. Thanks to Mark for his questions, editing of the interview, and permission to republish it here.

I also want to thank Becca Brown, Michele Boyd, and David Hoffman, who also provided suggestions during the Advisory Board meeting. I interviewed several people in industry, government, and academia, who gave generously of their time and expertise. They included Isaac Berzin (GreenFuel Technologies), Dick Post (Lawrence Livermore National Laboratory), Joel Schindall (Massachusetts Institute of Technology), John Morris (Sunlight Direct, Inc.), Michael Winkler (Schatz Energy Research Center, Humboldt State University), Randy Zwetzig (Ice Energy), and John Miller (Maxwell Technologies). John Miller also provided comments on the draft report. Melissa Kemp provided me with data and insights into the practical world of residential solar photovoltaic system costs and installations. And thanks also to Scott Powell for his volunteer proofreading. I would also like to thank all those who generously gave permission to use their images. They are credited at the appropriate locations in the text.

This project was something of a research adventure and the IEER staff played a critical role. Lois Chalmers was, as usual, a careful fact-checker and compiled the bibliography and found reference materials for me. Annie Makhijani compiled the energy use and supply data, checked my calculations, researched several topics, including biofuels, and did many of the graphics. Betsy Thurlow-Shields typeset the manuscript with Ken Shields' help. Natasha Makhijani researched documents on energy efficiency and compiled them into an easy list. She also proofread the manuscript, copy-edited it, and helped prepare the index. Brice Smith checked my energy demand spreadsheets and some of the financial calculations. Dr. Pushpa Mehta did the original painting reproduced on the cover especially for this book. Her artistry and generosity (she donated the work) are deeply appreciated.

It has been a special pleasure to work with Julie Enszer, Executive Director of NPRI. She has managed this project with insight and shepherded it along with care. For their support of this project, NPRI and IEER wish to thank The Park Foundation, The Lear Family Foundation, The Lintilhac Foundation, and many individual donors who wish to remain anonymous. Grateful appreciation is extended to all who supported this work.

Arjun Makhijani

PREFACE

A three-fold global energy crisis has emerged since the 1970s; it is now acute on all fronts.

1. Severe climate change, caused mainly by emissions of carbon dioxide from fossil fuel burning and associated emissions of other greenhouse gases;
2. The security of oil supplies, given the political and military turmoil in much of the oil exporting world, centered in the Persian Gulf region;
3. Nuclear weapons proliferation and its potential connections to the spread of nuclear energy to address climate change.

These issues are intimately connected. Oil is a leading source of global and U.S. carbon dioxide (CO_2) emissions as well as a principal source of local air pollution, and often the main one in cities. Concerns about the insecurity of oil supply are not new – they were expressed as long ago as 1952 by the Paley Commission,[1] when the United States was just turning from an oil exporter to an oil importer. To complicate matters, many, including some environmentalists, now propose that nuclear power should be one of the sources of energy used to reduce carbon dioxide emissions. The U.S. energy legislation of 2005 provides significant subsidies, not only for renewable energy sources, but also for new nuclear power plants.[2] But nuclear power and nuclear weapons proliferation are quite entangled with one another.

This report is not about the tangle of these difficult problems, but about a central, indeed indispensable, part of the solution – greatly reducing U.S. emissions from fossil fuel burning, which constituted 84 percent of U.S. greenhouse gas emissions in 2004. Its focus is to assess the feasibility of a zero-CO_2 economy in the United States and to lay out a roadmap to achieve that as early as is technically and economically practical, without resort to nuclear power. This preface lays out the reasoning for that framework and discusses the scope of the report.

A. Climate Change

The end of 2006 and the start of 2007 saw a flurry of initiatives from business, Congress, and the Bush administration,[3] on energy and climate change that seems to provide some hope the United States, by far the richest country in the world and the largest emitter of greenhouse gases, will begin to take national action. Many states, local governments, some corporations, many non-government groups, scientific panels, as well as many European countries had begun to take action years ago.

Action is surely necessary. The evidence of serious climate change, induced mainly by anthropogenic emissions of greenhouse gases, is now overwhelming; it need not be recounted here in detail, since this report is devoted to solutions. A few bullet points will suffice:

- Glaciers are melting across the world.
- Arctic ice is disappearing at a much faster rate than estimated just a few years ago – fast enough for the U.S. Fish and Wildlife Service to propose putting polar bears on the endangered species list.[4]
- CO_2 is a greenhouse gas and has increased by more than one-third in the last century and a half, due to human emissions.[5]
- Millions of acres of Alaskan forests are dying of insect infestations because the summers are longer and much warmer.[6]
- The Siberian permafrost is beginning to melt, raising the possibility that large amounts of methane now immobilized in the permafrost as methane hydrates would be released into the atmosphere.[7] (Methane is the primary chemical component of natural gas.) Such releases could suddenly aggravate existing trends and make temperature increases and sea-level rise greater and faster than now estimated. Even a one or two foot average sea-level rise would cause severe harm to tens of millions of people living in coastal areas around the world, from Florida to Bangladesh to the small island countries.
- Evidence of more frequent extreme climatic events is mounting. It is still difficult and controversial to try to pin a single extreme event, such as a hurricane, on climate change. But there is enough cumulative evidence to indicate that suffering and grievous damage of the type experienced in 2005 by the people of New Orleans and other parts of the U.S. Gulf Coast may become more frequent. The economic consequences will be long lasting. The population of New Orleans has not recovered. The poor and African Americans continue to be disproportionately affected, raising larger questions about society's ability to equitably handle more frequent serious climate-induced disruptions.

As of early 2007, the atmospheric concentration of CO_2 is over 380 parts per million (ppm).[8] Some ecosystems are already being extensively damaged, notably coral reefs.[9] The consequences that are unfolding from the tropics to the tundra do not depend on additional increases, which will only make the problem worse. The most recent work of the Intergovernmental Panel on Climate Change estimates that the cumulative CO_2-equivalent must be within the 445 to 490 parts per million range in order to limit the mean global temperature rise to 2.0 to 2.4 degrees Celsius (3.6 to 4.3 degrees Fahrenheit).[10] According to the Stern Review, at that level, we risk the "possible onset of collapse of part or all of the Amazonian rainforest,"[11] which has been called the lungs of the planet. The estimated effects at various levels of CO_2-equivalent concentrations of greenhouse gases are shown in Figure P-1, reproduced from the Stern Review.

The Intergovernmental Panel on Climate Change (IPCC) estimates that it will be necessary to reduce global carbon dioxide emissions by 50 to 85 percent relative to 2000 by 2050 in order to limit the temperature rise to less than 2 to 2.4 degrees Celsius.[12] With a 50 percent reduction, the IPCC estimates only a 15 percent chance of limiting the temperature rise to this range; with 85 percent CO_2 emissions reduction, the IPCC estimated that there would be an 85 percent chance of achieving the temperature limitation goal. Relatively simple calculations show that if global emissions are allocated according to even minimal norms of equity and the requirements of the United Nations Framework Convention on Climate Change, a near-total elimination of emissions from fossil fuels will be required in the United States (see Chapter 1).

B. Nuclear Power and Nuclear Weapons Proliferation

The connection of nuclear power to potential nuclear weapons proliferation has been recognized as a potential problem from early in the nuclear age. Yet, the urgency of the buildup of greenhouse gases is such that nuclear power is being promoted in quarters other than the nuclear industry as a part of the solution to greatly reducing greenhouse gas emissions.

IEER has addressed the inadvisability of such a course in the past, including recently in great detail, in a book by Dr. Brice Smith entitled *Insurmountable Risks: The Dangers of Using Nuclear Power to Combat Global Climate Change*.[13] Nonetheless, given the importance of the nuclear power debate and its security significance, the arguments are summarized in Appendix A of this book. In brief, the core arguments relate to:

- nuclear non-proliferation (and the connections between nuclear power and nuclear weapons technologies and infrastructure);
- the risks arising from severe accidents on the scale of the 1986 Chernobyl accident. Though the probabilities of an accident vary from one reactor to the next and are likely much lower in the United States than in the former Soviet Union (given historical data), accidents on the scale of Chernobyl could occur in all commercial reactor designs;
- the nuclear waste problem, which has not been solved so far in any country; The significant long-term health, environmental, and safety problems associated with spent fuel or high level waste disposal continue to bedevil nuclear power and make its future uncertain. It should be noted in this context that official assessments of the risk of harm from exposure to radiation continue to increase;[14]
- the high financial risks of nuclear power, including long-lead times and uncertainties relating to high level nuclear waste disposal, including the costs of repositories;

- the insurance problem. The damage from severe accidents has always been officially assessed as so severe that the nuclear industry continues to rely essentially completely on government-provided insurance, which itself is capped at a level far lower than official accident damage estimates.

It is strange that more than half a century after the then-Chairman of the Atomic Energy Commission, Lewis Strauss, proclaimed that nuclear power would be "too cheap to meter," the industry is still turning to the government for loan guarantees. But it should not be a surprise, since the original "too cheap to meter" campaign was part of a global propaganda campaign designed to make the U.S. atom look peaceful following the U.S. and Soviet tests of thermonuclear weapons.[15]

Further, the Bush administration is jointly promoting a scheme with Russia that would deprive parties in good standing under the Nuclear Non-Proliferation Treaty (NPT) their right to acquire commercial nuclear power technology. Article IV of the NPT actually states that it is an "inalienable right." But the administration's "Global Nuclear Energy Partnership" proposes to restrict commercial uranium enrichment and plutonium separation to the countries that already have it.[16] It is also a transparent attempt to change the Nuclear Non-Proliferation Treaty without going through the bother of working with the signatories to amend it.[17] This undermines the treaty and non-proliferation generally.

Uranium enrichment is at the center of U.S.-Iranian nuclear tensions. Iran claims it is pursuing commercial nuclear power; the United States believes it is acquiring nuclear weapons capability. In reality, the two are compatible statements – and that is the core of the problem. Building large numbers of nuclear plants across the world will multiply the need for commercial uranium enrichment plants. It is unlikely that countries will voluntarily give up their right under the NPT to acquire them.

Already, a number of developments in the world, including the above mentioned concerns about Iran, as well as the failure to achieve progress towards a nuclear weapons free zone in the Middle East, envisioned by the parties to the NPT at the time of its permanent extension in 1995, have intensified interest in acquiring nuclear power infrastructure in the region. For instance, at its 27th Summit, the Supreme Council of the Gulf Cooperation Council, consisting of the United Arab Emirates, Bahrain, Saudi Arabia, Oman, Qatar, and Kuwait, announced its intent to pursue civilian nuclear power technology, with an unmistakable link to nuclear weapons developments in the region. The remarks of the Saudi Foreign Minister on this topic are reported in the following news story:

> The leaders of Saudi Arabia, Bahrain, Kuwait, Oman, Qatar, and the United Arab Emirates called for a peaceful settlement of the conflict over Iran's nuclear program, and demanded that Israel, the only country in the Middle East believed to have nuclear weapons, join the nuclear Non-Proliferation Treaty.

Speaking to reporters after the summit, Foreign Minister Prince Saud Al-Faisal said the GCC states' intention to pursue civilian nuclear technology was not a "threat" to anyone. "We are announcing our intention to pursue the ownership of nuclear technology for peaceful (purposes)," he said.

"It is not a threat. It is an announcement so that there will be no misinterpretation of what we are doing. We are not doing this secretly. We are doing it openly," he said.

"We want no bombs. Our policy is to have a region free of weapons of mass destruction," the prince added. "This is why we call on Israel to renounce (nuclear weapons)." The "original sin" was from Israel as it established a nuclear reactor with the only purpose of producing nuclear weapons, Prince Saud said.[18]

This is a recipe for an intensification of problems both in the oil sector and in nuclear proliferation. The time for preaching temperance from a barstool is over. The twentieth century saw countries slowly struggle for freedom from domination. Unfortunately as part of that process, they also viewed the world powers refusing to give up their own nuclear weapons, even though the latter retained unquestioned superiority in conventional weaponry and power. The best way to approach the problem of non-proliferation is for the United States to undo what it began with Atoms for Peace and replace it with energy for peace. This book shows it is possible to have a secure and economical energy system without the headaches and risks of nuclear power. Why would one want to expand its role in an already insecure world?

For the record, we are not opposed to all nuclear technology or even all nuclear power technology. Nuclear fission has been a problem, but certain approaches to nuclear fusion, such as the proton-lithium reaction, could result in excellent power sources, if they could be made to work. Unfortunately, nuclear fusion, whose scientific feasibility as a power source remains to be established, is too far off to help with the problem of abating CO_2 emissions. Hence it is not considered in this report.

It should also be noted that infrastructure for regulatory, safety, and training needs must be maintained for existing nuclear power plants until they are phased out. Even after that, the problem of spent fuel management and disposal will be with us for many years. But the bottom line has been clear for some time. To attempt to solve the problem of climate change by resorting to reliance on nuclear power would be to exchange one serious problem for another when there is no need to do so. This roadmap, therefore, seeks to lay out a course for a zero-CO_2 economy without resort to nuclear power. At the same time, it is also clear that nuclear power supplies too large a portion of U.S. electricity to be switched off quickly. Hence, the approach taken here is a phase-out of nuclear power plants as their licenses expire. This is a normative assumption, and the actual course will depend on the specific phase-out policy that is adopted, and the phase-out duration may be shorter or longer than that modeled here.

C. Oil

The use of oil is responsible for about 44 percent of U.S. fossil-fuel-related CO_2 emissions. Currently, U.S. requirements are just over 20 million barrels per day, about 60 percent of it being imported.[19] Whatever the reasons for the origins of the Iraq War, it now appears to be tangled up with concerns about the security of oil supply from the Persian Gulf.[20] Former Secretary of State Henry Kissinger noted emphatically in an op ed piece in the *Washington Post* that

> American forces…are in Iraq not as a favor to its government or as a reward for its conduct. They are there as an expression of American national interest to prevent the Iranian combination of imperialism and fundamentalist ideology from dominating a region on which the energy supplies of the industrial democracies depend.[21]

The Iraq Study Group put it less bluntly, but part of its message was the same.[22] The direct costs to the United States of the Iraq war are running at $100 billion per year – roughly $100 per barrel of oil imported by the United States from the Persian Gulf.[23] The human cost in lives of Iraqis and of U.S. and allied soldiers and other personnel is incalculable.

Oil and democracy have never mixed in the Middle East. Its very map and political arrangements were created by the West, notably by the British and the French, in the wake of the collapse of the Ottoman Empire after World War I, with an eye on oil.[24] Side by side with the technological brilliance that has resulted in a vast river of oil flowing from the depths under turbulent oceans and forbidding desert sands, oil has gone hand in hand with war, violence, intrigue, coups, counter-coups, and revolutions.[25] Now, it is tangled up with the terrorism and the War on Terror that the United States undertook in the wake of the attacks on September 11, 2001.

A flourishing U.S. economy that has vastly lower CO_2 emissions than at present is necessary – based on considerations of global climate change alone. But it is also indicated by the need for disentanglement of U.S. economic well-being from oil. Such a course would produce a situation in which the political and developmental interests of the people of the Middle East could be disconnected from the Western need for – or, as President Bush said in his 2006 State of the Union speech, "addiction" to – oil.[26]

D. Lifestyles and Values

The analysis in this book does not address lifestyles and values as they relate to energy. That omission has nothing to do with my assessment of the importance of the topic. Rather, it has to do with a practical consideration. My goal was to assess the technical and economic feasibility of a U.S. economy with neither nuclear power nor CO_2 emissions. This can be done in a most straightforward way by using standard economic assumptions about future sizes of homes and offices, numbers of personal vehicles, and overall income and expenditure in society. It so happens that the use of energy is so inefficient that a several-fold growth is possible in gross domestic product without any growth in energy use and even while energy use declines. For instance, it is possible to design homes with available technology and architectural concepts that use just one-tenth the energy per square foot as is typical at present. Similar economies are possible in personal vehicles and in the commercial sector. Our approach enables the technological, economic, and policy recommendations developed here to be compared to others that are part of the present climate change debate. It is therefore not necessary to the objective of this study to address the issues of lifestyles and values, though, of course, that does not diminish the importance of the topic.

A large number of other questions, including environmental and health questions, associated with an ever increasing flow of materials through society, are also important. For instance, the mining of copper, gold, titanium, tantalum, and other minerals on ever increasing scales, the making of large amounts of chemicals, and other similar economic activities create environmental and health problems that are far beyond the energy use involved. Mining also often contributes to regional and global inequities, whereby certain regions become suppliers of specific raw materials while other regions and people become the main consumers.

Finding better approaches to meeting the material needs of a comfortable life to which essentially all people aspire is critical to environmental protection but beyond the scope of this book, except for the energy aspect of the issue. But it is clear that such approaches are needed, if only to enable economic development to meet the needs of much of the world where a majority of people are still poor, and where millions of children go hungry to bed, which is often the floor of a mud hut.

Beyond the matter of better technical means, there is the question of how much material throughput the world can sustain. That issue is also beyond the scope of this study. But it is clearly important in a world of eight to ten billion people, who are acquiring the means to live well. For the first time in the history of civilization (societies ruled from cities), a world in which all people can realistically aspire to achieve a comfortable life appears to be a real possibility.

The history of development shows that the norms for the "good life" are set by the wealthy. In that context, it appears necessary to develop the notion of "enough." Such a notion is not contrary to the pursuit of happiness, in the felicitous phrase of the Declaration of Independence. Rather, research shows that once poverty has been overcome, money seems to make little difference to happiness.[27]

The problem of how a change in values might occur to a long-term sustainable pattern that includes economic life broadly is a complex one. Specific changes in economic culture can occur rapidly, as for instance, has happened in many urban areas with recycling. Separating trash into recyclable and non-recyclable parts was not considered very practical in the United States just two decades ago. But it is now the norm. This indicates that similar changes could also occur in personal habits and tastes in relation to broader choices, including the way we use energy, the settings of our thermostats, the size of our homes and cars, etc.[28] It is obviously desirable; but when and how it might occur is difficult to predict and quantify, which is one of the reasons it is not part of the analytical framework of this book.

E. Conclusions

The power of setting a goal of a zero-CO_2 economy should not be underestimated. A U.S. economy that is in a ferment of innovation and investment in efficiency and new energy sources and technologies will spur the world energy economy in the same direction far more powerfully than can now be imagined. Even a single, short paragraph in President Bush's 2007 State of the Union message about climate change reverberated around the world.[29] His promise at the G8 summit at Heiligendamm, Germany, in June 2007, that the United States would seriously consider at least a 50 percent cut in greenhouse gas emissions by 2050[30] has even bigger implications. It is functionally equivalent to a zero-CO_2 emissions economy, defined as being within a few percent on either side of complete elimination (see Chapter 1). More than 100 percent reduction would mean removal of some of the CO_2 that has already been emitted from the atmosphere. This may become necessary should climate change turn out to be more severe than now estimated.

The goal of zero-CO_2 emissions does not mean that other greenhouse gas emissions should not be addressed. They should be; in many cases large reductions can be achieved rapidly in these other areas. It makes sense to reduce such emissions along with reducing CO_2 emissions.[31] But the size of the fossil fuel contribution to greenhouse gas emissions in the U.S. picture is so large that any overall goal of greenhouse gas emissions reductions translates directly into about the same percentage goal for reduction in CO_2 emissions from fossil fuels.

A new determination in Congress, a greatly expanded leadership at the state level, the immense success of *Inconvenient Truth*, the documentary on climate change featuring former Vice President Al Gore, who has recently called for a 90 percent reduction in greenhouse gas emissions by developed countries,[32] and a remarkable and possibly historic statement calling for a 60 to 80 percent reduction in greenhouse gas emissions issued by the U.S. Climate Action Partnership are among the many signs that a moment of decision on action at the federal level on climate change is at hand or at least near in the United States.

The present movement towards action on climate change seems analogous to the 1985-1987 period, when environmentalists, scientists, corporations, the federal government, and other governments arrived at an agreement on ozone layer protection that pointed at first to a large (50 percent) reduction in emissions of chlorofluorocarbons. The agreement expanded rapidly towards a complete elimination of CFC emissions. There were those who feared that a rapid phase-out of ozone depleting compounds would send humanity back to the caves without refrigerators or air conditioners, but once the key players decided it was time, the changes were as remarkable as they were rapid.

My hope – and I know it is Helen Caldicott's as well – is that this report will provide the occasion for a national debate on setting a goal of eliminating CO_2 emissions for the U.S. economy as rapidly as is economically sensible without recourse to nuclear power. It is also intended as a stepwise but flexible technical and economic guide for the actions that are needed in the next two decades to set the United States on such a course. Helen and I also thought that it would help that debate if the project were to have a diverse and experienced Advisory Board to help shape the outline and review the draft report.

Arjun Makhijani
Takoma Park, Maryland
July 2007

Figure P-1. Stabilization Levels and Probability Ranges for Temperature Increases

The figure below illustrates the types of impacts that could be experienced as the world comes into equilibrium with more greenhouse gasses. The top panel shows the range of temperatures projected at stabilisation levels between 400ppm and 750ppm CO_2e at equilibrium. The solid horizontal lines indicate the 5 – 95% range based on climate sensitivity estimates from the IPCC 2001[2] and a recent Hadley Centre ensemble study[3]. The vertical line indicates the mean of the 50th percentile point. The dashed lines show the 5 – 95% range based on eleven recent studies[4]. The bottom panel illustrates the range of impacts expected at different levels of warming. The relationship between global average temperature changes and regional climate changes is very uncertain, especially with regard to changes in precipitation... This figure shows potential changes based on current scientific literature.

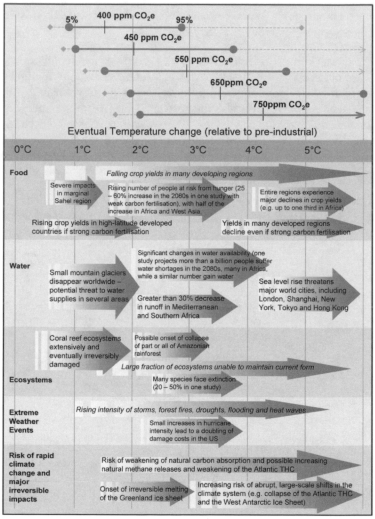

[2] Wigley and Raper 2001. [3] Murphy et al. 2004. [4] Meinshausen 2006

Source: Stern Review 2006, Executive Summary, Figure 2 (page v). Crown copyright material is reproduced with the permission of the Controller of HMSO and the Queens Printer for Scotland.

CHAPTER 1: SETTING THE STAGE

A. The Need for a Zero-CO_2 Economy in the United States

At the June 2007, G8 summit on Heiligendamm, Germany, the heads of state, including President Bush, made a commitment on climate change that implies drastic changes in the U.S. energy economy:

> Taking into account the scientific knowledge as represented in the recent IPCC reports, global greenhouse gas emissions must stop rising, followed by substantial global emission reductions. In setting a global goal for emissions reductions in the process we have agreed today involving all major emitters, we will consider seriously the decisions made by the European Union, Canada and Japan which include at least a halving of global emissions by 2050. We commit to achieving these goals and invite the major emerging economies to join us in this endeavour.[1]

The commitment was rather more vague than sought by the European Union, especially Germany's Chancellor Angela Merkel (who is also a physicist). The EU has the goal of limiting temperature rise to 2 to 2.4 degrees celsius, which implies reducing CO_2 emissions globally by at least 50 to 85 percent by 2050 (see below). But the statement was a radical departure for the Bush administration, which in its first year went back on its campaign statement that it would reduce CO_2 emissions, among other pollutants, from power plants.[2] Until 2007, it even showed a reluctance to acknowledge the seriousness or the urgency of the problem of human-induced climate change.

Global greenhouse gas emissions are a mix of emissions from fossil fuel use (55 percent) and other sources, such as methane emissions from landfills, pipelines, and agriculture (16 percent), nitrous oxide emissions from fertilizer use (9 percent), CO_2 emissions from forest burning and other land use changes (19 percent), and emissions of certain organic compounds known as halocarbons (1 percent).[3]

The situation for the United States is somewhat different in that a far larger proportion – 84 percent – of greenhouse gas emissions are due to CO_2[4] – almost all of it from fossil fuel use. Hence, any overall commitment for a reduction of greenhouse gas emissions will translate almost directly into a requirement for about the same reduction of CO_2 emissions from fossil fuel use.

Halving CO_2 and other greenhouse gas emissions would mean considerably larger cuts for Western countries, most of all the United States, which has the largest emissions. This is because developing countries will likely insist, at least, on an equal per-capita global norm, given historical inequities, even if it is not part of a formal agreement. Their arguments are straightforward and compelling:

- The vast majority of the increase in CO_2 concentration from the pre-industrial level of about 280 parts per million to about 380 parts per million in 2005 was due to the burning of fossil fuels in the West.
- The consumption of commercial energy in developing countries per person is far lower today, in part due to their long domination by the West, which began to be reversed only in the course of the twentieth century. The economies of many developing countries, especially China and India, which together have almost two-fifths of the world's population, are growing rapidly. Any arrangements that institutionalize material inequalities between developing countries and the West are very unlikely to be politically acceptable.
- China, India, and other developing countries are becoming the industrial manufacturing centers of the world. The Chinese have recently pointed out that much of the greenhouse gas emissions in China are actually attributable to exports consumed in the West.[5]
- Without the larger developing countries, such as China, India, Brazil, Mexico, and South Africa, in the dialogue there is little hope of actually achieving the needed reductions of global greenhouse gas emissions by mid-century.

A per-capita norm is therefore the minimum that would likely be needed for a global agreement to significantly reduce greenhouse gas emissions by 2050.

According to the Intergovernmental Panel on Climate Change, the likely range of CO_2 emissions reductions required by 2050 relative to the year 2000 for this goal is 50 to 85 percent in CO_2 emissions.[6] At the lower end of this range, a reduction of about 88 percent would be required in U.S. CO_2 emissions. At the higher end of this range, the U.S. reduction would have to be about 96 percent.[7] For the United States this translates directly into approximately the same reductions of CO_2 from the energy sector.[8] These figures are based on a per-capita norm.

Former Vice President Al Gore has called for a 90 percent cut in "global warming pollution...in developed countries." [9] Since the per person emissions in Europe and Japan are considerably lower than in the United States, this would amount to a reduction of about 95 percent for the United States.[10] But he has specified a framework for reductions that would imply an even greater reduction

– an essentially zero-CO_2 economy in the United States, Western Europe, and Japan. That is because his argument for CO_2 reductions goes beyond a per-capita allocation norm:

> A new [climate] treaty will still have differentiated commitments, of course; countries will be asked to meet different requirements based upon their historical share or contribution to the problem and their relative ability to carry the burden of change. This precedent is well established in international law, and there is no other way to do it.
>
> There are some who will try to pervert this precedent and use xenophobia or nativist arguments to say that every country should be held to the same standard. But should countries with one fifth our gross domestic product – countries that contributed almost nothing in the past to the creation of this crisis – really carry the same load as the United States?[11]

The most directly applicable international law is the United Nations Framework Convention on Climate Change (UNFCCC), which was ratified by the United States in 1992. It notes both the historical disparities in creating the problem as well as the present inequalities. The parties to the treaty noted that

> ...the largest share of historical and current global emissions of greenhouse gases has originated in developed countries, that per-capita emissions in developing countries are still relatively low and that the share of global emissions originating in developing countries will grow to meet their social and development needs...[12]

As a result, the UNFCCC places a greater responsibility on the developed countries for a reduction of emissions:

> The Parties should protect the climate system for the benefit of present and future generations of humankind, on the basis of equity and in accordance with their common but differentiated responsibilities and respective capabilities. Accordingly, the developed country Parties should take the lead in combating climate change and the adverse effects thereof.[13]

An equal per-capita norm is a minimal equity requirement of the UNFCCC. In sum, the demands of averting the worst effects of climate change and considerations related to global politics and international law combine to mean that the United States will likely have to eliminate 95 percent or more of its energy-related CO_2 emissions by the middle of the century. This is the definition of a zero-CO_2 economy discussed in the preface of this book. In point of fact, the practical actions that need to be taken to reduce emissions by 90 percent or more are along the same lines as those needed for a 100 percent elimination of CO_2 emissions. The sooner we prepare for and act to achieve a zero-CO_2 economy, the smaller will be the cost of the transition. One reason is that the less time we have to achieve this goal, the higher the fraction of expensive and less commercialized technologies that will have to be deployed to get there.

B. Historical Overview

Before the first energy crisis in 1973, it was generally accepted that growth in energy use and economic growth, as expressed by Gross Domestic Product (GDP), went hand in hand. In that year, in the midst of a period of rising demand, a political-military crisis in the Middle East enabled the Organization of Petroleum Exporting Countries (OPEC) to suddenly raise prices. At the same time, in October 1973, the Arab members of OPEC imposed an oil embargo on the United States and its Western European allies and Japan. Multinational oil companies were able to manage the global supply so as to keep the United States and other affected countries provided with oil (though not without some disruption and confusion). But the price increases and embargo caused the United States and Europe to take a fresh look at energy and, not least, at the assumption that energy demand growth and GDP growth were destined to be in lockstep.

The Ford Foundation's Energy Policy Project, headed by S. David Freeman,[14] was in the midst of producing technical scenarios and economic assessments that showed that the United States had wide latitude in choosing its energy future. Depending on the energy policy adopted, energy growth could continue in lockstep with economic growth ("business-as-usual scenario"), with attendant environmental and security problems, including growing dependence on imported oil, or modest energy growth ("technical fix scenario"), or even zero energy growth ("zero energy growth scenario") – the latter after a modest period (about ten years) of adjustment. As it turned out, the economic and political shock of rising energy prices and the oil embargo led the United States government, private industry, and not a few states, California being the first, to adopt energy policies and practices that transitioned to the new mode of economic growth without energy growth by the mid-1970s.[15]

Figure 1-1 shows the historical energy growth in the United States since 1949 and the clear, sharp break that occurred in 1973. The decline in energy use in the immediate aftermath was partly due to a recession, but economic growth resumed in the mid-1970s without energy growth (on average) until the mid-1980s. The economic-energy relationship overall and the relationship of energy sources to fossil fuel sources is shown in Figure 1-2.

After a decline in the immediate post-World War II decade, the energy required to produce a dollar of GDP stayed approximately constant overall until 1973 (with compensating variations within the period). Since 1973, there has been a steady decline, steep at first, in the period up to the mid-1980s, and then at a lower rate until the early part of the 21st century, but still much different than the period prior to 1973. As a result, in the year 2000, the energy required to produce a unit of GDP was about 55 percent of that in the mid-1950s. We note here that the period from 1982 onwards was characterized by falling petroleum prices and by a laissez-faire attitude to energy policy at a national level.

Figure 1-1: Historical U.S. Energy Consumption, by End Use Sector (Quadrillion Btu perYear)

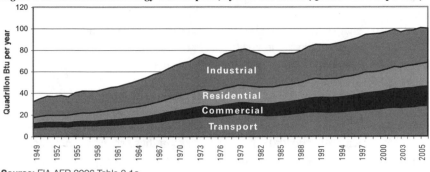

Source: EIA AER 2006 Table 2.1a

The decline in energy/GDP ratios, was reflected in the reduction of CO_2 emissions per dollar of GDP. In fact, the carbon/GDP ratio declined slightly faster than the energy/GDP ratio, notably in the 1950s and 1960s, reflecting the relative increase of the use of natural gas in the U.S. economy.

The decline in the carbon intensity of the U.S. economy was not reflected in a marked decline in the relative carbon dependence of the U.S. economy for a variety of reasons, including a continued reliance on coal for electricity generation and on oil for transportation. In other words, even as carbon emissions per unit of GDP declined, the dependence of the United States on fossil fuels as a proportion of its energy supply has not changed much since 1973. Hydroelectric power did not grow much, while nuclear power supplies only about eight percent of total energy use.[16] A central result has been the increasing dependence on imported oil, from about one-third of demand in the early 1970s to about 60 percent in recent years.[17]

Figure 1-2: Energy, GDP, and Fossil Fuel Relationships: History and Official Projections

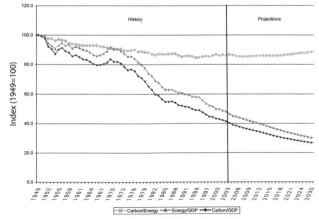

Courtesy of the Energy Information Administration of the United States Department of Energy

Even with a resumption of energy growth since the mid-1980s, the business-as-usual picture does not resemble the pre-1973 picture:

- Industrial energy use stayed about the same between 1973 and 2004, but the value of industrial production has more than doubled.[18]
- The ratio of energy demand growth to GDP growth has declined from about 0.9 in the mid-1950s-1973 period to about 0.5 by the year 2000 (See Figure 1-2). As in the 1973-1985 period, this increase in efficiency has been driven partly by price and partly by regulations.
- Residential, commercial, and transportation energy use has driven up energy use. Between 1995 and 2004 the growth rates in these sectors were 1.35 percent, 1.88 percent, and 1.60 percent respectively.[19]

In effect, "business-as-usual" in the industrial sector has meant economic growth without energy growth for over three decades. A part of this is may be due to the migration of energy intensive industries to countries with cheaper energy supplies. But a central factor has been an increase in efficiency of energy use in industry. Historical data for industrial energy use are shown in Figure 1-3.

Figure 1-3: Industrial Energy Use – Historical Data

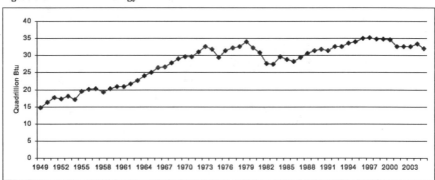

Source: EIA AER 2006 Table 2.1d

The overall trend to declining requirements of energy per unit of GDP is only partly due to prices. The decline in the use of energy per dollar of GDP has continued even through periods of declining energy, and especially petroleum, prices since 1973. The consistent trend, through both rising and falling prices, is largely due to

- Continued increases in industrial energy efficiency (in terms of energy input per dollar of output)
- Federal and state efficiency standards for appliances[20]
- Mileage standards for passenger vehicles that created very large energy efficiency increases in the first two decades after 1973.[21]

Figures 1-4, 1-5, and 1-6 show historical oil, electricity, and natural gas prices in constant 2000 dollars, respectively.

Figure 1-4: Historical Crude Oil Refiner Acquisition Costs, in Constant 2000 Dollars per Barrel

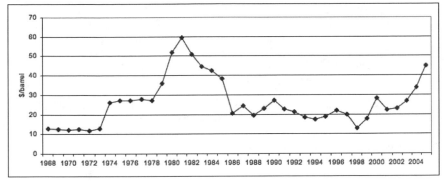

Source: EIA AER 2006 Table 5.21

Figure 1-5: Historical Average Retail Electricity Prices, in Constant 2000 cents per Kilowatt Hour, Including Taxes

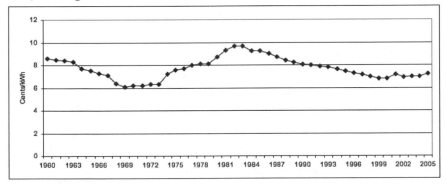

Source: EIA AER 2006 Table 8.10

Figure 1-6: Historical Natural Gas Prices by Sector, in Constant 2000 Dollars per Thousand Cubic Feet

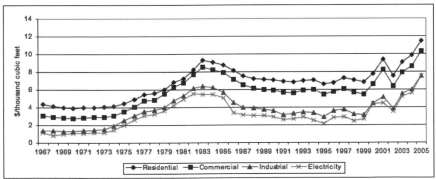

Source: EIA AER 2006 Table 6.8

The overall effects of the changes on the economy as a whole, as well as the energy sector, have been dramatic. Rosenfeld and McAuliffe have summarized the net effects by hypothesizing what might have been under "business-as-usual," i.e., a continuation of pre-1973 trends compared to the actual result, since 1973:

1. Under "Business as Usual," US primary energy demand could have been 170 Quads by 2005 rather than the Actual 100 Quads

2. Energy expenditures in 2005 could have been $1.7 trillion rather than $1.0 Trillion. The savings are on the order of $700 billion. To put this into perspective, U.S. energy purchases totaled about $ 1 trillion in 2005 out of the GDP of 11.7 trillion (nominal dollars or $10.8 trillion in chained 2000 dollars).

3. We only had to meet 25 Quads of increased demand for primary energy, not 95 Quads (the difference between 170 Quads and 75 quads in 1973). The remaining 70 Quads were avoided. To be able to deliver an additional 25 Quads, hundreds of power plants were built, refineries upgraded and expanded, new tankers constructed, pipelines and transmission facilities added and coal, natural gas and petroleum combusted. Alternately, to avoid 70 Quads we drastically changed our energy policies, invested in more efficient buildings and appliances, altered our transportation fleet to be much more fuel efficient, developed new and ingenious products and processes, and responded to increasing prices in many other ways.[22]

However, the State of California has done much better than the national norm. Figure 1-7 shows the evolution of per person electricity use in California since 1960. In 1976, the national figure was only about 15 percent greater than that of California. By the turn of the century, it was 70 percent greater. California's milder climate cannot explain most of the trend since the relative climate situation is approximately the same today as it was three decades ago. It is the more active approach to energy policy that California has taken that is mainly responsible for the difference.

Figure 1-7: California Electricity Use Trends Compared to the United States

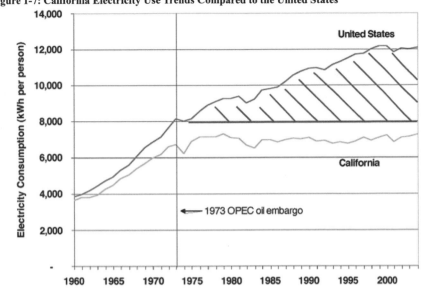

Per Capita Electricity Consumption in California and the U.S.

Source: Chang, Rosenfeld, and McAuliffe 2007 Figure 7 (page 13)

The hatched area in Figure 1-7 provides an approximate idea of the excess U.S. electricity consumption per person since 1973 relative to California. It represents about ten billion metric tons of CO_2 extra emissions in the United States relative to California policies.

The relative unimportance of climate is also indicated by the fact that the states with the lowest energy use per unit Gross State Product (GSP) are not necessarily the ones with the mildest climate. Figure 1-8 shows CO_2 emissions per person by state. CO_2 emissions are a good proxy for energy use, since about 86 percent of energy use involves burning of fossil fuels.[23] Leaving aside the District of Columbia because it is a city, the other states with low per-capita emissions have widely varying climates.

Figure 1-8: CO$_2$ Emissions per Person by State, 1999

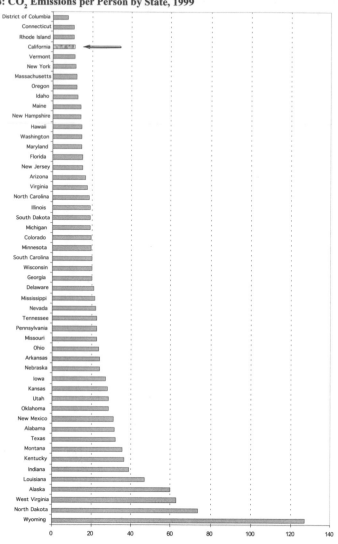

Source: Bemis and Allen 2005 Figure 8 (page 18)

Finally, Figure 1-9 shows the metric tons of CO$_2$ emissions per thousand dollars of gross state product. It is evident that per-capita emissions are more linked to the structure of the economy than to the weather. For instance, mining and agricultural states, like Wyoming, West Virginia, or Kansas, tend to have higher per-capita emissions than service and manufacturing states like the New England or mid-Atlantic states or California, even though the heating and cooling requirements among the latter group of states is quite variable. Some states like Wyoming also have mine-mouth coal-fired plants for exporting electricity out of state.

Figure 1-9: CO$_2$ Emissions per Gross State Product by State: 1999 (in Metric Tons of CO$_2$ per Thousand Dollars)

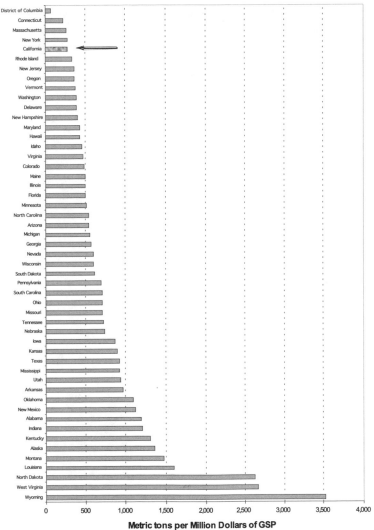

Metric tons per Million Dollars of GSP

Source: Bemis and Allen 2005 Figure 9 (page 19)

The factors that go into the energy structure of an economy are obviously quite complex in their technical detail. But it is clear that from a macro-economic point of view, market factors and regulatory policies can have and have had a fundamental impact on the structure and amount of energy consumption per person or per unit of economic output. The reason is not far to seek. Existing efficiencies of energy use are quite low by the criterion of how much of the available energy is actually applied to the task at hand. After all, except for sunshine and food, energy is not a need in itself (though sometimes it is still discussed that way). It is the services that energy provides that are important.[24] For instance, when we flip a light switch that turns on an incandescent light bulb, only about 1 percent of the fuel input into electricity generation shows up as visible light. High-efficiency compact fluorescent lamps reduce energy consumption by about a factor four while providing approximately the same visible light output. As another example, photoelectric switches that turn off outdoor lights in the daytime or motion detectors that turn off lights when rooms are not occupied do not change the utility provided by energy use to people, but reduce energy use and greenhouse gas emissions.

Many of these changes, including adopting the use of motion detectors, photoelectric switches, efficient electric motors for industrial applications, and compact fluorescent lamps can be accomplished more economically than the present high energy use, high emissions approach. A rather dramatic example of a change brought about by energy efficiency standards for appliances is provided by refrigerators. In 1973, the electricity use per cubic foot of an average refrigerator freezer was about 100 kilowatt hour, electrical. California enacted standards in 1978 that were then tightened. The federal standards went into effect in 1990 and tightened subsequently. The typical refrigerator in 2001 consumed about only about a fifth as much per cubic foot,[25] despite having more features. *Moreover, real prices of refrigerators have come down significantly in the same period, despite larger size. Between 1987, when federal standards were enacted, and 2002, the unit value of a refrigerator fell from about $575 to just over $400.*[26]

As a final example, consider the efficiency of personal passenger vehicles. Only about 15 percent of the energy contained in petroleum actually winds up as mechanical energy that moves the car or SUV from one place to another.[27] Moreover, the "payload" in the car, the weight of the passengers, is about seven percent of the weight of the vehicle, using the average vehicle weight of 3,240 pounds[28] and occupancy of 1.64 person-miles per vehicle mile.[29] Hence, the actual energy used to provide the utility for which the car is designed to move people from one place to another is typically about one percent.

The use of lighter, stronger materials that provide safety similar to heavier vehicles, regenerative braking, automatic engine cutoff when the car is stopped, more efficient engines, and efficient electric cars are all approaches that can greatly improve the efficiency of passenger transport. Excellent public transport, which makes for more livable cities, might increase GDP and improve the environment in a variety of ways, while at the same time decreasing energy use by reducing the need for personal vehicles for commuting, shopping, etc. Many of these approaches have been tried on various scales. The goal here is to explore a more efficient energy economy that is set in the technical context of zero-CO_2 emissions in the supply sector. The social goal is that this transition should be accomplished with justice for the affected workers and communities.

Plan of the Book

A combination of efficiency increases and changes in the sources of energy supply will be needed to achieve a zero-CO_2 economy. We first provide an overview of the macroeconomic assumptions for the energy economy in Chapter 2. This chapter also includes the economic assumptions regarding energy prices and the implicit price on carbon dioxide emissions under various circumstances. Energy supply and storage technologies and their possible evolution in the next decade or two are explored in Chapter 3.

Chapter 4 sets forth the demand-side scenario for each broad consuming sector, along with the technology assumptions, that provides the basis for the analysis of options for a zero-CO_2 energy supply. When all is said and done a large supply of energy will be required for a U.S. economy that is three times larger than today, even with great improvements in energy efficiency. Chapter 5 describes a reference scenario for a zero-CO_2 emissions economy. Chapter 6 describes variations on the reference scenario. The objectives of describing a reference scenario and possible variations are to

- Demonstrate that a zero-CO_2 economy, without recourse to nuclear power, is possible within a few decades.
- Explore the land-use implications of a large-scale reliance on biofuels.
- Explore alternative approaches to meeting the requirements of critical and difficult sectors such as aircraft fuel.
- Explore possible alternative paths that would make the transition faster, more economical, and/or more desirable from other economic, environmental, and security standpoints than the reference scenario.

Chapter 7 discusses the policy framework at the federal and state levels as well as actions that can be taken at the private level – whether corporate or individual drawing on existing examples. Finally, Chapter 8 sets forth a roadmap for a zero-CO_2 economy without nuclear power, with goals and policies that need to be taken and alternatives that need to be pursued. Note that electricity generation costs are based on 2002-2004 data. Costs of most sources except solar and some new technologies have been rising, which will make efficiency and solar energy more attractive than some of the estimates in this book. The plan here is to develop an approach that will have flexibility built into it. The aim of the roadmap is not so much to look into an energy crystal ball and foretell the exact route all the way to a zero-CO_2 emissions economy but to set forth a technical and policy approach that can deal with uncertainties and setbacks. The principal technical approach is to develop backup technologies and multiple approaches to the same result. In that case, if some of the advanced technologies that now appear promising falter, there will be others to take their place. Chapter 9 summarizes the main findings and recommendations.

CHAPTER 2: BROAD ENERGY AND ECONOMIC CONSIDERATIONS

Since the mid-1990s, the efficiency of energy use per unit of GDP has been increasing at about two percent per year on average.[1] On this basis, a three percent annual GDP growth would result in energy growth of about one percent per year. This scenario, which we might call business-as-usual in the present context – that is, assuming no dramatic changes in energy prices or policies, would result in an increase in energy use from about 100 quadrillion Btu in 2004[2] to about 160 quadrillion Btu in 2050 (all figures are rounded). Energy use actually declined slightly in 2006 to below the level in 2004.

Official energy projections corresponding to expected trends under prevailing conditions, that is, corresponding to business-as-usual trends, prepared by the Energy Information Administration (EIA), go only to 2030. The demand projection is shown in Figure 2-1 and the supply projection is shown in Figure 2-2.

Figure 2-1: EIA Projection of Energy Demand, by End Use Sector to 2030

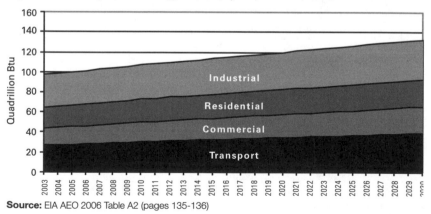

Source: EIA AEO 2006 Table A2 (pages 135-136)

Carbon-Free and Nuclear-Free | *A Roadmap for U.S. Energy Policy*

Oil and coal, the main sources of CO_2 emissions in the United States, are projected to grow the fastest. Nuclear energy, often presented as being the solution or at least a major part of the solution to global warming, is officially projected to decline in share from eight percent today to less than seven percent in 2030.

Figure 2-2: EIA Projections for Energy Supply, by Fuel, in Quadrillion Btu

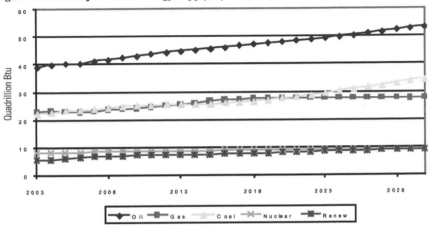

Source: EIA AEO 2006 Table A2 (pages 135-136)

Note: EIA AEO 2006 does not give the breakdown for renewable energies, but says that the contribution is mostly from hydroelectricity and biomass (wood and ethanol), not wind and solar energy.

In this book, we use present energy use along with the economic assumptions in the EIA projections to create the reference energy and economic scenario that is needed to explore approaches to a zero-CO_2 economy.

Figure 2-3 shows the floor space projections for the residential and commercial sectors and Figure 2-4 shows the projections for the transportation sector in terms of the demand for services, based on present trends of square feet per house or office, number of homes, growth in passenger miles traveled by road and air, etc. These projections are extended to 2050, based on the reference conditions underlying the EIA projections to 2030 in Figures 2-1 and 2-2 above.

Figure 2-3: Residential and Commercial Sectors, Projections of Floor Space, in Billion Square Feet

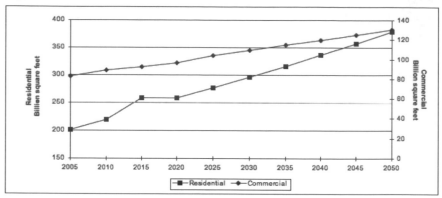

Sources: Commercial: EERE 2006 Table 2.2.1 (page 2-5) and Residential: EIA AEO Assumptions 2006, page 23 and EERE 2006 Table 2.1.1 (page 2-1) EIA AEO gives an average square footage for 2001 and 2030. We have interpolated the values for the years in between and multiplied them with the number of households listed in EERE 2006. The values after 2025 for commercial area and after 2030 for residential area were extrapolated.

Figure 2-4: EIA Transportation Projections, in Billion Vehicle Miles Traveled (for Light-Duty Vehicles) or Billion Seat Miles Available (Aircraft)

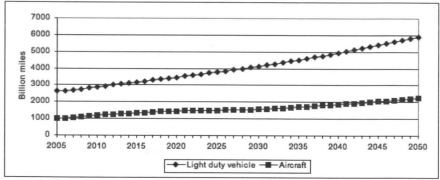

Source: EIA AEO 2006 Table A7 (pages 145-146) up to 2030, projected thereafter by IEER.
Note: Light duty vehicles are defined as weighing less than 8,500 pounds.

While it is possible to construct zero-CO_2 scenarios at various levels of overall demand (including energy conversion losses in electricity production), even for those above the level of about 100 quadrillion Btu in 2004, the pressure on resources, notably land, could be serious (see Chapters 5 and 6). Moreover, the economics of attempting to do so would also be dubious at best and, more realistically, poor. Even at present prices, there are plenty of foregone opportunities for energy efficiency investments due to a variety of factors. For instance, developers of residential and commercial real estate generally do not pay the utility bills. Automobile manufacturers do not pay the fuel bills. These disconnects create economic inefficiencies as well as pollution. They mean that policies that ensure more cost-effective and environmentally sound results, while allowing markets to function in terms of allocating investments, are essential.

Cost effectiveness will be set in the context of policies that are aimed at reducing and then eliminating CO_2 emissions. For instance, a system in which large users of energy must buy allowances for emitting CO_2 will increase the effective price of fossil fuels, making both renewable energy sources and efficiency measures relatively more attractive. Carbon taxes could, in theory, accomplish the same purpose (see Chapter 7). For instance, energy use in industry has stayed constant over more than three decades without carbon taxes and with fluctuating energy prices. With higher fossil fuel costs in the form of a price on CO_2 emissions, it is reasonable to expect that industrial energy use would decline somewhat – possibly at a rate of one or two percent per year.[3]

As will be discussed in more detail, the opportunities in the transportation, commercial, and residential sectors for economic implementation of energy efficiency are substantial. For instance, well-insulated homes designed to capture solar heat passively – that is, in their structures – can eliminate most of the space heating requirements under most circumstances prevailing in the United States. And near-term technology will allow far greater efficiencies in all sectors. For instance, all-electric cars are now being made with a new generation of lithium-ion batteries in which the carbon has been eliminated for safety reasons and which can be charged in ten to fifteen minutes at a gas station-like service stop. First generation all-electric cars and pickup trucks made with lithium-ion batteries can go 3.3 to 5 miles on a single kilowatt hour of electricity. Plug-in hybrids can get 70 to 100 miles per gallon with an input of just over 0.1 kWh of electricity.

The analysis of energy efficiency potential in this report indicates that instead of requiring one percent energy growth for three percent economic growth (the approximate business-as-usual case), the same economic growth can be accomplished with an absolute reduction of about one percent in delivered energy use per year. (Delivered energy excludes electricity losses in electricity generation and other losses incurred in the production of the energy supply; it includes only the energy as consumed at the point of end use.) Such an approach would make a transition to a low or zero-CO_2 economy much more manageable both for creating the supply from renewable sources and for transitioning to a better balance between supply and efficiency than has been characteristic of the U.S. economy in the past. With a special emphasis on the transportation sector efficiency, it would also alleviate the security concerns now associated with the large-scale of oil imports on which the U.S. economy is now so dependent.

A one percent decrease in delivered energy use per year means approximately two percent per year overall improvement in efficiency compared to recent trends (discussed in Chapter 1 above). This would mean that instead of delivered energy growing from about 75 quadrillion Btu per year in 2005 to about 120 quadrillion Btu in 2050, it would decline to between 40 and 50 quadrillion Btu. This is shown in this study to be eminently feasible, largely with existing technology.

Significantly greater efficiencies are possible in many areas but they have not been assumed in the reference scenario (See Chapter 6).

A. Analysis of Energy Prices and Implicit CO_2 Prices

Any substantial reduction of CO_2 emissions implies some price that would be attached to CO_2 emissions. For instance, the cost of coal-fired generation from a new pulverized coal-fired power plant is about 4 cents per kWh.[4] But these plants, of course, emit the most CO_2 of any type of large-scale power plant– about 950 grams per kWh.[5] Policies to reduce and eventually eliminate CO_2 emissions would therefore effectively attach a cost to the fossil fuel user for emitting the CO_2 that was at, or just above, the cost of reducing the marginal emission at any particular stage. That is, if the user faces the prospect of paying a price for a CO_2 emission allowance just greater than the cost of eliminating the emissions of CO_2, investments would gravitate to the necessary areas to reduce the emissions. The cost can be added in various ways, by imposing taxes, regulations, or caps on emissions implemented through auctions of CO_2 emission allowances (a "hard cap" on emissions that would decline in quantity each year). These approaches are discussed in Chapter 7.

In this report, however, we seek to achieve multiple objectives: eliminating CO_2 emissions and nuclear power in the same process and also ensuring the reliability of liquid fuel supplies, which today are mainly in the form of petroleum.

The marginal cost of reducing CO_2 emissions varies a great deal according to the application. Sometimes, the implicit CO_2 price may even be negative. In other words, the cost of doing things with lower CO_2 emissions may be lower than the methods used at present. Combined heat and power generation in a part of the commercial sector (large buildings, for instance) provides an example in many circumstances.[6]

The exercise here, in the context of a goal of zero-CO_2 emissions, is to assess the implicit CO_2 price of eliminating essentially all the CO_2 from a given sector on the understanding that the price of CO_2 emissions allowances would rise to this level in the last stages of CO_2 emissions elimination (assuming orderly and efficient markets in CO_2 emission allowances).

1. Implicit CO_2 Price in the Electricity Sector

Let us first consider direct elimination of CO_2 from a coal-fired power plant in its simplest conceptual form. To do this, we try to estimate a market price that would result in a steady reduction of CO_2 from the electricity sector, recognizing that different technologies would come into play at different stages.

The most straightforward approach to estimating a long-term price for reduction of CO_2 emissions from coal-fired power plants is to consider the cost of preventing CO_2 emissions from such a plant. A commonly proposed way for doing this is to use a coal gasification system combined with a power plant. The system is called the Integrated Coal Gasification Combined Cycle (IGCC) power plant. The CO_2 generated by the combustion process is captured, rather than being emitted to the atmosphere. It is then piped to a location where it can be injected into a deep geologic system, where it would be expected to remain for thousands of years. The entire system is called carbon capture and sequestration (CCS). This system has been much studied and is being developed because of the extensive use of coal in the electricity generation systems of the United States, China, Russia, India, and other countries.

The main difficulty lies in estimating a cost of sequestering carbon dioxide successfully for thousands of years in deep geologic formations. Injection of carbon dioxide into oil and gas reservoirs for stimulating production has been done commercially; sequestration of CO_2 in geologic formations on a limited basis has also been demonstrated.

However, there is also some uncertainty as to the long-term success of sequestration. With many reservoirs required for large-scale application of the technology, it is possible that one of them could fail and suddenly emit a large amount of carbon dioxide. Since CO_2 is denser than air, it would hug the ground, possibly asphyxiating a nearby population. This has occurred in the case of a natural venting of CO_2 from a lake in western Africa in 1986.[7] The question of liability associated with such venting from CO_2 sequestration is an important one both from the point of view of safety of nearby populations and for financial risk. The process of safely siting CO_2 repositories and the cost and availability of insurance are still open questions, especially given the long time frames involved.[8] There is also some uncertainty associated with what it might cost to make sure that sequestration has low leakage rates over thousands of years.[9] In other words, though CO_2 injection into geologic reservoirs has been demonstrated, there are still outstanding issues in applying it to the vast amounts of CO_2 that are generated by coal-fired power plants and in ensuring that the CO_2 remains sequestered for very long periods of time.

Present estimates of cost are made on the basis of rather limited experience relative to requirements of sequestering billions of metric tons of CO_2 each year if large-scale use of coal continues. Nonetheless, the available data provide a useful benchmark in attempting to estimate how much it would cost to prevent CO_2 emissions compared to operating pulverized coal-fired power plants. The estimated costs have a wide range, which provides one indication of the uncertainties. Overall, the added costs of an IGCC plant and the capture, transport, and sequestration of CO_2 have been variously estimated as being between 1 cent and 4.2 cents per kWh compared to a pulverized coal plant with no CO_2 emission controls.

For the purposes of this report, we will assume a cost range of 1 to 4 cents per kWh for carbon capture and sequestration, in order to develop an implicit CO_2 price. The term "CO_2 price" is a theoretical price that would have to be charged to a power plant owner in order to induce the installation of equipment to prevent the CO_2 emissions. Of course, this does not ensure that the equipment will be installed; rather it provides a way of comparing the costs of different approaches of avoiding CO_2 emissions. Different policy approaches to actually accomplish that have their own advantages and disadvantages. These are discussed in Chapter 7.

If the price of an emissions allowance for a metric ton of CO_2 emitted is $10, a power generating company would, in theory, be willing to spend almost that much to capture and sequester CO_2. At about 35 percent generation efficiency, the added cost would amount to about 1 cent per kWh. Since the cost range for IGCC with carbon capture and sequestration is estimated to be in the range of 1 to 4 cents per kWh, the CO_2 price that would induce an investment in CCS would be $10 to $40 per metric ton.

We can also develop a price to be imputed to CO_2 (that electricity generators using coal would pay) by comparing the cost of replacing electricity from coal with electricity from nuclear power. The base case estimate range provided in the MIT study published in 2003 was 6.7 to 7 cents per kWh, or nearly 3 cents more than coal.[10] The assumptions underlying this study are somewhat optimistic, given the experience of building nuclear power plants in the United States in the 1980s and 1990s. For instance, it assumes a construction time of six years and an overnight capital cost (assuming zero construction time) of $2,000 per kilowatt. The CEO of Duke Energy, which owns nuclear power plants and advocates building more, stated in 2007 that the cost was likely to be more in the $2,500 to $2,600 range.[11] Further, there are large uncertainties in relation to the cost of spent fuel management. With the one investigated disposal location facing delays and questions about its licensability (Yucca Mountain in Nevada), it is unclear what the costs of deep geologic disposal might be. The Bush administration is pursuing a reprocessing initiative for commercial spent fuel. If this is actually pursued as the main disposal path, it could add at least 2 cents per kWh or more to nuclear electricity generation costs. Two cents per kWh is the estimated added cost of the world's largest program (as implemented by France) to reprocess spent fuel and to use the separated plutonium as a fuel in reactors.[12]

A realistic range of nuclear power costs, not taking into account insurance subsidies and uncertainties relating to proliferation, severe accidents, or prolonged construction delays, is that it would be 2 to 5 cents per kWh higher than the cost of coal-fired power plants without CO_2 capture and sequestration. It corresponds to a CO_2 price of $20 to $50 per metric ton of CO_2 emissions.[13]

There are options for reducing CO_2 emissions that can be achieved at lower costs. For instance, if time-of-use pricing is permitted – that is, if the price recovered during peak and intermediate hours is relatively high – off-peak wind energy can be priced at 2 to 3 cents per kWh. Under these circumstances, the early reductions in CO_2 emissions from coal-fired power plants could be achieved by purchasing off-peak wind power and reducing output from coal-fired power plants, which have off-peak costs of about 2 cents per kWh. The implicit cost range for avoiding CO_2 emissions in this case is zero to $10 per metric ton of CO_2. However wind energy has added transmission and infrastructure costs. Adding these costs yields an estimate of $5 to $15 per metric ton of CO_2 for using off-peak wind to displace coal.

For the initial tranches of CO_2 reductions, it is possible that an emerging technology may provide an opportunity for negative CO_2 costs – that is, if the costs are roughly as projected by the developer, it would be possible to reduce CO_2 emissions commercially, even in the absence of climate change considerations. Technology to capture CO_2 from power plant effluent gases in microalgae grown in plastic tubes exposed to sunlight was recently demonstrated on a significant scale at a 20 megawatt (MW) natural gas-fired cogeneration plant at MIT. According to the leader of the technical team that developed the technology, Isaac Berzin, the algae can be profitably converted to biofuels (biodiesel and ethanol) so long as the price of petroleum stays above about $30 a barrel. The approach is in the engineering demonstration phase. A 0.3 acre plant has been built in cooperation with Arizona Public Service.[14] The performance of the plant at MIT in terms of CO_2 capture efficiency has been independently confirmed. The technology has not yet been commercialized and the developer's cost estimates remain to be demonstrated both for microalgae and liquid fuel production.

This cost structure must be reevaluated for a higher penetration of renewables, when the intermittency of wind and solar energy becomes more of a concern. Some portion of the intermittency problem in wind can be addressed by geographical diversity. Another very important portion can be addressed by coordinating and optimizing the capacity of central station solar power plants built in sunny areas, such as the Southwest and parts of the West, with large-scale wind farm installations. Since the weather is more predictable from the standpoint of day-ahead planning for central station solar power plants, standby capacity requirements can be minimized. Further optimization can be achieved by taking advantage of the fact that, in many areas, the wind blows preferentially in the evening and night hours, thus complementing solar energy during the daytime.

Finally, solar thermal plants can also be built with a few hours of storage to supply the peak demand in the early evening hours (see Chapter 3). Still, one can safely assume that a considerable reserve capacity in some form will be required at high penetration levels of wind and solar.

The most readily available, large-scale reserve of electrical power generation capacity is combined cycle natural gas plants.[15] A vast expansion of such plants began in the 1990s making them economically attractive. The capacity was built to operate economically at natural gas prices of $2 to $3 per million Btu, which were the prevalent prices in the electricity sector through almost the entire 1990s (see Figure 1-6). Construction of such plants continued into the first years of the present decade, when natural gas prices fluctuated a great deal. They have stayed above $4 for the electric generation sector since about 2003 and were about $8 per million Btu in 2005.[16] The net summer capacity for natural gas-fired power plants in 2005 was 383,000 megawatts.[17] The high price of natural gas has meant that at the present time the capacity utilization of these plants is very low – in 2005, the average capacity factor was only about 22.6 percent.[18]

At $8 per million Btu, the fuel cost alone for a typical combined cycle power plant is about 5.6 cents per kWh.[19] After adding a variable maintenance cost of about 0.5 cents per kWh, the off-peak avoided cost is about 6 cents per kWh (rounded). This is greater than the cost of new wind energy capacity of about 5 cents per kWh.[20] At natural gas prices of about $6.50 per million Btu, natural gas power combined cycle power plants can be idled and kept on standby at zero added cost to provide electricity when wind farms cannot meet demand. There is an implicit net zero-CO_2 price at $6.50 per million Btu of natural gas since at that price the marginal operating cost of the natural gas plant is about equal to that of new wind capacity. At natural gas prices greater than $6.50 per million Btu there would be a net reduction in overall generation cost if combined cycle capacity is idled in favor of wind. This means that at current prices of about $8 per million Btu, CO_2 emission reductions can be achieved by using wind to displace combined cycle and single stage turbine capacity with a net economic benefit to consumers in the form of lower electricity prices.

It is possible, of course, that natural gas prices will again decline below $6.50 per million Btu. This would create a positive implied CO_2 price. At $4 per million Btu, which is approximately the cost of marginal supply (imported liquid natural gas, or LNG), the off-peak marginal cost of a combined cycle plant is 3.3 cents per kWh. With wind at about 5 cents per kWh, there is then a 1.7 cent per kWh differential. This corresponds to a CO_2 price of about $46 per metric ton. At $5 per million Btu, the implicit CO_2 price is about $26 per metric ton. Combining the best wind sites with combined cycle natural gas standby will likely be economical at $5 per million Btu of natural gas or more at an implied CO_2 price that is zero or negative (that is, a net reduction in cost would be achieved). There are also other options for standby capacity for renewables in the long-term.

Compressed air storage could be used, for instance (see Chapter 3). Another example is the potential for using plug-in hybrids or all-electric cars using new designs of lithium-ion batteries in a vehicle-to-grid (V2G) mode, where electricity flows from the cars back to the grid at certain times of the day. Such cars are expected to be economical at a battery cost of about $200 per kWh of storage. One battery design has been tested in the laboratory over more than 10,000 charging and discharging cycles (see Chapter 3). The collective installed power of automobiles is vastly greater than that of the electric power system. It should be possible to provide backup power using vehicle-to-grid at a modest cost, using only vehicle fleets (such as corporate or government fleets under contract for such services) and parking structures in the commercial sector.

If battery life proves to extend in practice to over 10,000 charging cycles, then the marginal cost of the V2G would be very low. It would essentially equal the electricity losses in the battery, which are low. This is because over a ten or twelve year vehicle life, the expected number of charging cycles for motor vehicle operation itself would be far lower than 10,000. The main costs would be for the V2G infrastructure itself. One study of fuel cell vehicles estimated them to be about 0.5 cents per kWh for an operation involving 5,000 vehicles providing 10 kW each.[21] The energy-related costs would be those associated with the electricity losses in charging and discharging the battery, however, these are small.[22] Some rental charge would be paid to the vehicle owner and the owner of the docking station. If the battery depreciation is low, this cost could also be low. Assuming an overall added cost of 0.5 cents per kWh in this evaluation gives a total cost estimate about 1 cent per kWh. In other words it would cost $10 to reduce CO_2 emissions by one metric ton. Of course, this calculation is contingent upon the technology becoming economical in the coming years. However, in the context of the options for eliminating CO_2 from the electricity sector, it would not be needed for perhaps two decades, since other options to reduce CO_2 with present or near-present technology are available. We have used a cost estimate of less than $26 per metric ton for V2G to replace natural gas standby for wind.

In the near future, plug-in hybrids are a logical place to start building the infrastructure for efficient transportation and vehicle-to-grid experimentation. These are gasoline-electric cars that have extra batteries that store enough charge to enable much or most commuting on electricity only. Depending on the battery capacity, the liquid fuel efficiency is 70 to 100 miles per gallon. There is no real obstacle to commercialization of this technology. Efficiency standards set for the year 2020 should reflect this. And plug-in hybrids should become standard issue for federal government cars by 2015 (see Chapter 3 for more details and Chapter 7 for a policy discussion).

In sum, the short-term CO_2 emissions can be reduced from fossil fuel power plants at low cost – in the zero to $15 per metric ton of CO_2 range.[23] In the long-term a zero-CO_2 economy appears to imply a price of CO_2 of $10 to $40 per metric ton for a goal of eliminating it from the electricity sector. Given the financial, proliferation, and other risks associated with nuclear power (see Appendix A), it is difficult to justify reliance on nuclear power to reduce CO_2 emissions. Equal or lower cost solutions are either available or on the near-term (ten-year) horizon. The available data certainly do not justify providing subsidies to nuclear power plants to further climate change goals. On the contrary, the same money could be used to greater effect in other sectors.

Table 2-1: Summary of Costs for CO_2 Abatement (and Implicit Price of CO_2 Emission Allowances) - Electricity Sector

CO_2 source	Abatement Method	Phasing	Cost per metric ton CO_2, $	Comments
Pulverized coal	Off-peak wind energy	Short-term	$5 to $15	Based on off-peak marginal cost of coal
Pulverized coal	Capture in microalgae	Short-and-medium-term	Zero to negative	Assuming price of petroleum is >$30 per barrel
Pulverized coal	Wind power with natural gas standby	Medium-to-long-term	Negative to $46	High costs corresponds to a low natural gas price ($4 per million Btu)
Pulverized coal	Nuclear power	Medium-and-long-term	$20 to $30	Unlikely to be economical compared to wind with natural gas standby
Pulverized coal	IGCC with sequestration	Long-term	$10 to $40 or more	Many uncertainties in the estimate at present. Technology development remains.
Natural gas standby component of wind	Electric vehicle-to-grid	Long-term	Less than $26	Technology development remains. Estimate uncertain.

Notes:
1. Heat rate for pulverized coal = 10,000 Btu/kWh; for natural gas combined cycle = 7,000 Btu/kWh.
2. Wind-generated electricity costs = 5 cents/kWhe; pulverized coal = 4 cents per kWh; nuclear = 6 to 9 cents per kWh.
3. Natural gas prices between $4 and $8 per million Btu.
4. Petroleum costs $30 per barrel or more.
5. CO_2 costs associated with wind energy related items can be reduced by optimized deployment of solar and wind together (see Chapter 5)

2. CO_2 and Petroleum

Assessing the implicit price of CO_2 at which petroleum-related emissions would be eliminated is much more complex than the analysis for the electricity sector presented above for a variety of reasons:

- Unlike coal, almost all of which is used on a large-scale in electricity generation or industry, most petroleum is used in transportation in a manner that makes capture of the CO_2 practically impossible. Hence, no direct estimate of costs of CO_2 capture and sequestration is possible.

- The cost of producing oil in much of the world, notably the Persian Gulf region, is unconnected with its price. In the prolific oil fields of the region the cost is less than $3 per barrel,[24] while the price has fluctuated in the past decade between just over $12 and well over $70 per barrel.[25] The marginal cost of production from Canadian tar sands is about $30 to $35 per barrel, well below spot market oil prices since 2005.[26] Fluctuations in future prices based on non-economic security and political factors are still possible and may be considered likely.
- The indirect security costs of imported petroleum to the United States are high. If one is to take one's cue from Henry Kissinger, as quoted in the preface, then the need to continue a U.S. military involvement in Iraq is centered on protecting the flow of oil from the Persian Gulf region. In that case, the security cost of oil imported from the Persian Gulf by the United States amounts to about $100 per barrel. It is still about $22 per barrel if the cost of the war is spread out over all U.S. oil imports.[27]
- The net greenhouse gas reductions of ethanol made from corn, the largest alternative fuel in the United States, are small. Moreover, estimates vary considerably, making a net estimate of cost per unit reduction of equivalent CO_2 emissions very difficult. Whatever the exact figure, the cost would be very large because the net emission reduction is low, indicating that more efficient approaches need to be pursued.[28]

Security costs in the sense discussed here are distinct from any costs associated with reduction of CO_2 emissions. In theory, a security cost, distinct from a CO_2 reduction cost, should in some way be reflected in the price of petroleum and products derived from it. But how should such a security cost be calculated and how much should be attributed to petroleum? Answers to such questions are certain to be very controversial and difficult. It is unclear, for instance, whether the $100 billion per year being spent on the direct costs of the Iraq war should be attributed entirely to petroleum imports. That does not take other foreign policy goals into account. On the other hand, $100 billion per year represents only a very partial accounting of the total costs of the Iraq war. It does not include expenditures on the care of injured veterans, for instance.

We can approach the question of costs of reducing petroleum use and CO_2 emissions at the same time in a somewhat different way, at least for passenger vehicles. We will use a reference price range of $50 to $70 per barrel for petroleum here. This is above the marginal cost of $30 to $35 per barrel (from Canadian tar sands), which is the cost of extracting and producing the most expensive oil that is on the market today in significant quantities. The spot market price for crude oil over the past two years has been considerably over $50 per barrel and is about $70 per barrel at the time of this writing (early July 2007).[29] At $50 per barrel, the retail price of gasoline would be somewhat under $2 per gallon, including refining, retailing, and transportation costs, but not including taxes. With taxes, it would be about $2.25 to $2.50.[30] Using $2.25, the annual fuel cost of operating a typical 25 miles per gallon vehicle for 15,000 miles is $1,350. At $70 per barrel the price is closer to $3 per gallon, which gives an annual fuel cost of $1,800. If we add $0.50 per gallon for security costs, $0.50 for air pollution costs, and $0.50 for costs of avoiding CO_2 emissions, a reasonable overall working figure for social cost of fuel is about $4.50 per gallon. This gives an annual operating cost of $2,700.

We can now consider a reference vehicle used in this report for personal passenger transport and estimate what added costs can be paid for the vehicle at this price to eliminate gasoline use. Google is monitoring its plug-in hybrids for gasoline and electricity consumption. The average in early July 2007 was 73.5 miles per gallon and also uses 0.113 kWh per mile of electricity.[31] If it is mainly charged off-peak, the annual operating costs would be $564 to $717 (for $2.25 and $3 per gallon of gasoline). Using a discount rate of 7 percent over five years, typical of a car loan, an added cost of $3,310 to $4,560 for a plug-in hybrid can be accommodated without a change in overall operating costs relative to the average car. If the environmental and security costs are added, then an added cost of over $7,000 can be justified for a plug-in hybrid.

It is possible that the imputed price of CO_2 in the transportation sector could be very low. In the discussion on electricity above, we briefly discussed the capture of CO_2 from fossil fuel power plants in microalgae for the purpose of producing liquid fuels (biodiesel and ethanol) from it. Ethanol can be used as a feedstock for producing biobutanol, which is a direct gasoline substitute.[32] If the estimates made by Isaac Berzin, the Chief Technology Officer of GreenFuel are close to the mark, then liquid fuels could be economically produced if crude oil prices are above about $30 per barrel. Since this is about equal to or less than the marginal cost of oil production (from tar sands) of $30 to $35 per barrel, the imputed cost of CO_2 in this case would be zero or negative. At the present time, the overall system has not been demonstrated on a large-scale, so there is some uncertainty about cost estimates.

B. Defining "Zero-CO_2 Emissions"

As noted in the preface, the term "zero-CO_2 emissions" is not to be taken literally in the sense of eliminating the last ton of CO_2 emissions. A margin of a few percent either way would need to be preserved, especially when the zero-CO_2 target is connected with a particular date or narrow range of dates. We elaborate on this concept here.

It is possible that in some sectors the cost of eliminating fossil fuels may turn out to be high. For instance, aircraft can only be fueled with renewable energy sources in two ways, liquid fuels made from biomass or hydrogen made from renewable energy sources. Land constraints on the former may become important, especially if there are large demands for liquid and gaseous fuels in other sectors, such as cars and industrial feedstocks. A hydrogen-based air transportation sector is in the infancy of its development (though the technology has been shown to be feasible). Moreover, burning hydrogen creates water vapor, which acts as a greenhouse gas, especially if emitted at altitudes much above 30,000 feet (see Chapter 4). Hence, a considerable trade-off between economy, energy efficiency, and exchanging one greenhouse gas for another may face this sector. It is difficult to foresee how that might affect the price of biofuels or the price of the last five or ten million metric tons of CO_2 allowances for the commercial air transport sector. The approach in this report is to set forth options that can result in eliminating CO_2 emissions, but also preserve flexibility in the energy sector sufficient to prevent disruptions in the U.S. economy. The research for this study did not uncover any insuperable problems to actually eliminating all CO_2 emissions associated with the energy sector.

There is also the prospect that achieving zero-CO_2 emissions will not be enough, due to the accumulated impact of past emissions. At an atmospheric concentration of 380 parts per million of CO_2 today, there are already indications of serious climate change. Even if we reach zero-CO_2 emissions globally by mid-century, greenhouse gas concentrations are set to go beyond 450 parts per million CO_2 equivalent. In this context, it may well be necessary to go beyond zero-CO_2 emissions. This means we must make provision for technologies that could remove CO_2 from the atmosphere at reasonable costs.[33] What the extent of the need to go beyond zero-CO_2 emissions to negative CO_2 emissions (i.e., net removal of CO_2 from the atmosphere) will be for the United States is not now possible to foresee. This is especially so given that the first stage of the job – turning the economy around from a direction of increasing CO_2 emissions to one of decreasing CO_2 emissions – has barely begun. Hence, it is prudent to set a course that would aim for a zero-CO_2 economy, but also one that would allow for net removal of CO_2 from the atmosphere should it be deemed necessary.

In sum, the scenarios in this study are oriented to examining the feasibility of an actual zero-CO_2 economy, and to creating a roadmap for how it might be accomplished. So in the context of the technical analysis of the numbers in this report, zero-CO_2 is taken literally. However, in the context of the policies that are outlined, the term is regarded with more flexibility –"zero" is to within a few percent of present-day CO_2 emissions.

CHAPTER 3: TECHNOLOGIES—SUPPLY, STORAGE, AND CONVERSION

A large and fundamental transformation of the energy supply system will have to occur in the coming decades in order to transition to an economy with zero-CO_2 emissions without nuclear power. The division of investment resources between supply, storage, conversion (to electricity and/or hydrogen), and efficiency in utilization of energy will vary with policy and prices, but a basic reshaping of energy supply must take place. In this chapter, we will survey the energy sources that can provide the basis for such a transformation along with the conversion and storage technologies that are likely to be needed. Specifically, the configuration and roles of conversion and storage technologies in the electricity grid will be very different in a context where there are no fossil fuels or nuclear power. The grid itself will be much more a distributed grid, with generating plants of all scales contributing significant amounts, rather than one that depends almost wholly on central station power plants, which is the case at present. Further, with solar and wind energy playing very large roles, the role of storage and standby capacity will be more important than it is today.

This survey of technologies does not aim to be comprehensive. There is a tremendous ferment of innovation (literally and figuratively) in energy and it would take volumes to do technical justice to properly evaluate and compare the potential of the various ideas that are being developed. Even so, such a survey is likely to be quickly overtaken by events. The aim here is to present a sufficient technical evaluation of major energy supply sources and delineate the potential of each as it is best understood today so as to be able to create credible supply scenarios by combining them (Chapters 5 and 6). Some connection to the realities of the present demand structure are also needed, since not all energy sources can, at present, supply all demand sectors:

- Solid fuels – coal mainly – are used primarily in electricity generation and to a much lesser extent in industry (steel, cement, paper),

- Liquid fuels are used mainly in transportation and in industry, with feedstock use being a major application in the latter,
- Natural gas is used in the residential and commercial sectors mainly for space and water heating, for electricity generation, and for many applications in the industrial sector,
- Electricity is used widely in all sectors except transportation.

Table 3-1 shows the structure of energy supply in the United States, along with the main applications for each fuel in 2004. Table 3-2 shows a breakdown for natural gas use in 2004.

The connections of fuels to major end uses are not fixed, of course, but there is a considerable inertia in the system in that the utilization equipment, such as heating systems in homes and office buildings or boilers and process heat in industry, is structured to use certain fuels. Hence, the new supply sources also need to be evaluated for the kinds of demand they may satisfy and how the evolution of the demand sector may affect supply-side developments. Such considerations are left to Chapter 5, where a reference zero-CO_2 scenario is developed and to Chapter 6, where options for optimizing the system and providing flexibility and backup are discussed. These provide the basis for the policy considerations (Chapter 7) and the roadmap (Chapter 8).

Table 3-1: U.S. Energy Supply, 2004, in Billion Btu

Fuel	Billion Btu	Percent	Comments
Coal	22,603,933	22.5	Mainly for electricity generation
Gas	23,035,841	22.9	See Table 3-2
Oil	40,593,665	40.3	Mainly transportation and industry
Nuclear	8,221,985	8.2	Electricity generation
Hydro	2,690,078	2.7	Electricity generation
Renewable	3,529,674	3.5	Wood, geothermal, wind (electricity generation)
Total	100,675,176	100.0	

Source for the individual fuels: EIA AER 2006 Table 1.3

Table 3-2: Natural Gas Consumption in the United States, 2004

Sector	Percent
Industrial	37
Electricity	24
Residential	22
Commercial	14

Source: EIA AER 2006 Table 6.5

The major sources of renewable energy supply considered here are:

- Wind energy
- Solar energy, not including biofuels, but including solar photovoltaics and solar thermal power plants
- Solar energy in the form of biomass, including biofuels derived from it
- Direct hydrogen production from solar energy
- Hot rock geothermal energy
- Wave energy

We assume that hydroelectric resources will remain about the same as they are today.

The first four resources have the theoretical potential to supply the entire U.S. energy requirement. However, each faces certain constraints, such as intermittency with wind and solar, and land-area considerations with biofuels. In the case of use of solar energy for direct hydrogen production, a considerable amount of technological development remains to be done. It is included here because of its overall potential to transform the biofuels portion of a renewable energy structure in ways that would have a number of benefits compared to most biomass-based biofuels.

A. Wind Energy

Wind-generated electricity has been growing very rapidly in the last decade. Additions to capacity around the world far outstrip nuclear energy. In the United States, no new nuclear plants have been completed in many years and, despite much talk and expenditure, none have been ordered since 1978. The last order to be completed and commissioned was placed in October 1973. In contrast, wind capacity grew by about 2,700 MW in 2006 alone in the United States,[1] enough to supply the output of about one large nuclear power reactor. Similar additions to capacity are expected in the coming years. Figure 3-1 shows the Colorado Green Wind Farm, near Lamar, Colorado. (See color insert.)

Table 3-3 shows the wind energy potential in the top 20 states. It does not include offshore potential.

Table 3-3: Wind Energy Potential in the Top 20 Contiguous States, in Billion Kilowatt Hours/Year

State	Wind potential
North Dakota	1,210
Texas	1,190
Kansas	1,070
South Dakota	1,030
Montana	1,020
Nebraska	868
Wyoming	747
Oklahoma	725
Minnesota	657
Iowa	551
Colorado	481
New Mexico	435
Idaho	73
Michigan	65
New York	62
Illinois	61
California	59
Wisconsin	58
Maine	56
Missouri	52
Total	**10,470**
U.S. elec. generation, 2005	**4,000 (rounded)**
Potential percent of 2005 generation	**261 percent**
Wind energy generation, 2006	about 30 (0.7 percent)

Sources: AWEA 2006b; EIA AER 2006 Table 8.2a, AWEA 2007, and EIA AEO 2006 Table 16.
Note: For wind class category 3 and higher. Land use exclusions such as national parks, urban areas, etc., have been factored in to the estimate.

It is clear that overall potential is vast – over two-and-a-half times total U.S. electricity generation in the United States in 2005. The wind energy potential in *each one* of the top six states – North Dakota, Texas, Kansas, South Dakota, Montana, Nebraska – is greater than the total nuclear electricity generation from all 103 operating U.S. nuclear power plants. The wind energy resource is quite sufficient to supply the entire electricity requirement of the country for some time to come under any scenario, if total potential were the only consideration. Of course, it is not. Intermittency is a critical issue. Secondly, the geographic location of the wind resource is another potential constraint. It is concentrated in the Midwest and the Rocky Mountain states while the population of the United States is concentrated along the coasts. Figures 3-2(a) and 3-2(b) illustrate this issue; the former shows population density and the latter shows the map of wind energy.[2] (see color insert) Tapping into a large amount of the high-density land-based wind resource will require transmission infrastructure to take the electricity to transmission system hubs from where it would be taken to population centers. Transmission corridors exist going eastwards and westwards from the center of the country. But the wind resource is dispersed and it must be delivered to the hubs. Second, the capacity of some of the lines to carry the electricity would have to be expanded. The maps illustrate the importance of developing offshore wind energy resources, which are closer to the large population and electricity consumption centers of the United States.

One advantage of the geographic concentration of wind resources in the continental United States is that much of it is located in the Midwestern Farm Belt. Since crops can be planted and cattle can graze right up to the wind turbine towers, wind farms are quite compatible with growing crops and ranching. They can provide a reliable and steady source of income to farmers and ranchers, insulating them, to some extent, from the vagaries of commodity markets.

The largest single problem with wind energy is intermittency. This intermittency affects the system at many levels: short-term wind fluctuations, hourly or daily variations, and week-to-week and seasonal variations.

Figure 3-3 shows wind energy availability over a week compared to the fluctuations in electricity demand. Note that in this example, wind is frequently low at times of peak demand. Capacity of various types could be planned if wind could be accurately forecast. Day-ahead forecasts that are reasonably good and hour-ahead forecasts that are more accurate (on average) can be made, though there are times when the wind will be above or below those forecasts, occasionally by large amounts. The variability of wind energy therefore necessitates the addition of reserve capacity other than wind that can be tapped when the wind falls below the forecasted level over a period of hours or days. Electricity system planning takes place over various time intervals, with power plant availability being planned at all times from daily to seasonal.

Figure 3-3: Illustration of Wind Energy Variability

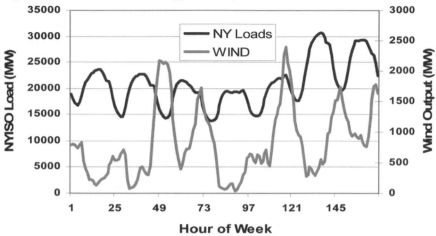

Hour of Week

Provided by the U.S. Department of Energy. **Source:** Parsons et al. 2006 Figure 5 (page 7)
Note: The wind capacity is shown on the right hand scale and does not contribute more than 10% of demand at the highest wind generation.

Besides the need for extra reserves, there are other costs of wind integration with electricity grids. Winds fluctuate over very short periods of time (seconds to minutes) creating disturbances in the system that could affect the stability of the frequency of the electricity supply. A constant frequency (in the United States, 60 cycles per second, called 60 hertz) is essential for much consuming equipment, such as clocks and computers and automated controls in industry dependent on electronic timing systems. The frequency of the electricity supply is therefore maintained within narrow limits at all times. The added cost of maintaining constant frequency as the proportion of wind energy in the system increases is called the regulation cost.

In between these two times scales (seconds to about a day) is the issue of load following. As we turn lights on and off and industries are brought on line or taken off, as millions of televisions are turned on in the evening when people return home from work, the electricity system must be able to follow the load and increase or decrease the output according to the demand. This is more complex if there is no actual control of the fuel supply that can change the output, which is the case with wind energy. It is analogous to a third party controlling the accelerator of a car.

These issues are managed by having some form of added reserve capacity and the reserves have to increase as the proportion of wind-supplied electricity increases. This is obviously an added cost that must be attributed to wind energy. It is the grid equivalent of having a battery storage for solar or wind energy in off-grid systems. Since loads can fluctuate rapidly over periods of minutes, every electricity system must have spinning reserve capacity – that is, capacity that is available whenever the demand goes up – somewhat like electricity "on tap." The additions to reserve capacity needed for maintaining the reliability of supply are a critical aspect of wind energy integration into electrical grids and represent part of the costs of this energy source. These costs are low when the proportion of wind-generated electricity is small, and tend to rise as that proportion increases.

Wind energy is now becoming a mature and very large industry. By the end of 2006, the total world wind energy capacity was over 74,000 MW – a capital investment worth about $100 billion. The worldwide additions to capacity in 2006 were about 15,000 MW – that is, the capacity grew about 25 percent in one year and is set to grow that much again in 2007. The United States' total capacity by December 2006 was 11,600 MW or 15.6 percent of the world total.[3]

A great deal of effort, study, and practical experience has gone into addressing problems such as wind integration to rather high levels of generation – up to about 20 percent – mainly in Europe (Denmark, Germany, Spain). Though the penetration of wind in the U.S. electricity market is still very low (about 0.7 percent of electricity generation), there have been many rigorous studies of wind integration costs. Overall, these have been assessed to be modest – in the range of 0.25 to 0.5 cents per kilowatt hour ($2.50 to $5 per megawatt-hour (MWh). For instance, the National Renewable Energy Laboratory published brief descriptions of several studies. One study in Minnesota found $4.60 per MWh was a conservative estimate of wind power integration cost at a level of 15 percent capacity:

> The costs of integrating 1,500 MW of wind generation into the Xcel North control area in 2010 are no higher than $4.60/MWh of wind generation and are dominated by costs incurred by Xcel Energy in the day-ahead time frame to accommodate the variability of wind generation and associated wind-generation forecast errors. The total costs include about $0.23/MWh resulting from an 8-MW increase in regulation requirements and $4.37/MWh resulting from scheduling and unit commitment costs. The study characterized these results as conservative, since improved strategies for short-term planning and scheduling and the full impact of new regional markets were not considered.[4]

Another study described the 300 MW pumped-storage (that is, the use of excess wind capacity to pump water from a low reservoir to a high reservoir) in Xcel's Colorado service territory. The water can then be run through an existing hydroelectric plant when the wind is not blowing. This smooths out some of the fluctuations in wind energy availability and reduces the costs of integration of wind into the grid. The cost reduction is dependent on the contribution of wind-generated electricity to the total. At a 10 percent level, the cost reduction estimated was $1.30/MWh.[5]

Development of wind resources in a manner that takes advantage of the large areas over which the resource is available would provide a great advantage in that it reduces the time when generation from wind energy is zero or very low. Studies have found that the costs of wind energy integration into the grid can be kept modest or small up to fairly high levels of penetration if geographic diversity is taken systematically into account as one design factor in the utilization of the resource.

A study commissioned by the Minnesota state legislature, published in November 2006, has examined this issue in considerable detail.[6] It found, for instance, that the ability to forecast available wind resources was considerably improved when the geographic diversity of the wind generation was increased. Hence, the dispersion of wind generation not only reduces the times for which no or low wind energy is available, it also improves the reliability of forecasting upon which reserve capacity requirements are based. Of course, this has a direct bearing on reducing the costs of integrating wind generation into the electricity grid. Table 3-4 shows that the reserve requirements for Minnesota's electricity system with 25 percent of the generation coming from wind would increase from 5 percent with no wind generation to just over 7 percent at the 25 percent level.

Table 3-4: Minnesota Reserve Requirements at Various Levels of Wind Generation

Reserve Category	Base		15% Wind		20% Wind		25% Wind	
	MW	%	MW	%	MW	%	MW	%
Regulating	137	0.65	149	0.71	153	0.73	157	0.75
Spinning	330	1.57	330	1.57	330	1.57	330	1.57
Non-Spin	330	1.57	330	1.57	330	1.57	330	1.57
Load Following	100	0.48	110	0.52	114	0.54	124	0.59
Operating Reserve Margin	152	0.73	310	1.48	408	1.94	538	2.56
Total Operating Reserves	1049	5.00	1229	5.86	1335	6.36	1479	7.05

Source: EnerNex 2006 Table 1 (page xvii)

A complementary approach, and one that would greatly increase geographic diversity, would be to develop offshore wind resources. This has been a topic of some controversy in the United States in a period when several European countries have developed significant offshore capacity and expertise. Offshore wind farms have other advantages besides being closer to large population centers. The wind over the oceans is steadier, providing for more reliable output and hence lower reserve requirements. A preliminary estimate of offshore U.S. wind energy resources (continental United States), excluding all areas within five nautical miles, two-thirds of the area between 5 and 20 nautical miles, and one-third of the area between 20 and 50 nautical miles is 908,000 megawatts of capacity.[7] This is sufficient to supply about 70 percent of U.S. generation in 2005.[8] Higher penetration of wind energy can and should be optimized with other renewable energy sources to take advantage of the diversity of supply and the greater ability of combinations of sources to more closely match demand. This is particularly true of wind and solar electricity. We are not aware of any thorough study (comparable to the many studies of wind integration) that has been done to examine the combinations of wind and solar electricity supply that could optimize cost and reduce requirements for reserve capacity.

Any large-scale development of wind resources or any other energy resource will have some environmental impact. Much of the focus for wind has been on bird kills, noise, and preservation of scenic values. The first two have largely been addressed by turbine design. The latter, of course, is a matter of one's aesthetics and how that competes with the need to reduce CO_2 emissions and with other available means to do so. Finally, very large-scale development of wind may also have climatic impacts that need to be more carefully studied. It has been postulated that wind power development may have adverse temperature change impacts, for instance. But such effects are not yet well-understood; indeed they are not yet amenable to reliable assessment. At levels 100 times today's level of wind penetration, at which level wind would supply a large fraction of the world's electricity requirements, the impacts may be somewhat negative to positive.[9] The reference scenario in this book envisages about a 20-fold increase in wind-generated electricity in the United States by about 2050 compared to 2006 but it would remain at less than 15 percent of total supply.

Small-scale wind turbines (a few hundred watts to 10 kW) are also available. These are considerably more expensive than large wind turbines and are used mostly for off-grid applications. There are also attempts to develop wind turbines for urban applications. This would work more like rooftop solar cells, with reverse metering. Such systems would be connected to the grid and feed into it or take energy from it depending on the wind level and the household demand. We will not consider these sources explicitly in this study, though they may become more important in present off-grid applications or, in the future, due to new designs and lower costs that would make them widely usable. The same considerations that apply to decentralized solar systems would also largely apply to decentralized grid-connected wind sources, though siting and some technical issues are likely to be more complex.

Large-scale wind energy development costs are about 4 cents per kilowatt hour at the very best sites to about 5 cents per kilowatt hour at very good sites, and about 6 cents per kilowatt hour at moderately good sites.[10] As discussed in Chapter 2, these costs are generally below the costs of new nuclear capacity. Wind energy is economical today. The main constraints lie in a lack of transmission infrastructure and an overall policy to reduce CO_2 emissions that would give rise to more rapid investments in this area.

B. Solar Electricity

The average solar energy incident on the continental United States is far greater than the wind energy potential. At about 5 kilowatt hours per square meter per day (annual average, 24 hours-per-day basis), the total is four thousand times the annual electricity generation in 2005.[11] Of course, only a small part of the area can be used and less than half of the incident energy is converted into usable electricity even under the best circumstances in a laboratory. But even at 20 percent efficiency and with one percent of the land area, the total potential for solar electricity generated by photovoltaic cells (solar PV) is about eight times the total U.S. electricity generation, and about three times greater than the wind energy potential shown in Table 3-3 above. Efficiencies of 40 percent have been demonstrated in concentrator solar cells in laboratory settings.[12] Twelve to eighteen percent is typical of non-concentrating solar PV silicon devices on the market today;[13] thin film solar cell efficiencies are typically several percent lower.

Unlike large-scale wind energy, solar PV is economical today in only some circumstances, but the economics of solar-generated electricity are improving rapidly. Typical retail costs for small-scale residential applications have been about $5 per peak watt for the solar cell module itself, besides installation costs. Total installed costs are often in the $8 to $9 per peak watt range.[14] These prices reflect silicon solar cells with traditional manufacturing technologies on a relatively small-scale backfitted onto existing homes. Prices have come down significantly in the last few years and continue to drop. For instance, according to the National Renewable Energy Laboratory, in 2004, installed costs for small-scale applications of thin film solar cells were about $6 per peak watt and up, of which about $3 was the solar cell cost.[15]

While the cost of solar PV installations is declining, it is still rather high, especially when it concerns traditional silicon solar cells and small-scale installations:

- the high price of crystalline silicon due to rapidly rising demand
- the small-scale of manufacture in typical solar cell plants, typically 20 to 30 MW of solar PV cells per year
- the high cost of traditional crystalline silicon manufacturing techniques
- the slow emergence of thin film solar cells, which do not use crystalline silicon, in large-scale manufacturing
- the deployment of solar PV in small-scale residential applications which are backfitted onto existing structures.

A number of factors are bringing down the costs of solar PV significantly. In the past year or two there have been significant new developments that would set a course for solar cells to have deployed costs of $2 or less per peak watt within a few years for intermediate- and large-scale applications (100 kw or more) and perhaps even for small-scale applications. It would take a considerable dissertation to go through the various developments, but the following list provides some indications of the basis for this conclusion:

- In June 2006, Nanosolar, a venture capital financed firm, secured $100 million in financing to build a 430 MW per year thin film solar PV factory in California. The scale of the manufacturing is large enough for the company to set a goal of competing with peak electricity generation costs. In a July 200 interview, the CEO of the company stated that volume manufacturing by 2008 would be the key to success in the industry and that Nanosolar would have certified solar panel "available in near-term 100MW volume at a fully-loaded cost point in the sixties [cents/Watt] or less so that one can profitably sell at a $.99/Watt wholesale price point."[16]
- First Solar, one of the larger solar PV manufacturers using thin film technology, announced that it had achieved a manufacturing cost as low as $1.25 per peak watt in its February 13, 2007, 8-K filing with the Securities and Exchange Commission. First Solar has signed contracts to supply 685 megawatts of solar PV to European clients for $1.28 billion, which is just under $1.90 per peak Watt. [17]
- A South African-German consortium that began building a thin film solar cell factory in Germany in 2006 announced anticipated costs of about one euro per peak watt[18] – about a factor of three to four less than present typical costs.
- A radically new manufacturing technique ("string-ribbon" technology) for polycrystalline silicon cells that draws strings of silicon through a silicon melt and produces very thin sheets cuts silicon requirements for solar cells by almost half, from over ten grams per watt for conventional ingot-based technology to six grams per watt. Further reductions in thickness are expected.[19]
- The first factory based on this technology, with a capacity of 15 MW of solar PV modules is operating in Marlboro, Massachusetts, and one with twice the capacity is operating in Thalheim, Germany.[20]
- The Department of Energy projects that annual manufacturing capacity of solar PV in the United States will increase almost twelve times in five years, from 240 megawatts per year in 2005 to 2,850 megawatts per year. It estimates that this expansion of capacity "put the U.S. industry on track to reduce the cost of electricity produced by PV from current levels of $0.18-$0.23 per kWh to $0.05 - $0.10 per kWh by 2015 – a price that is competitive in markets nationwide."[21]

To gain a perspective on these costs, the present electricity cost of new solar PV projects of intermediate or large-scale of about 20 cents per kWh about the same as that using a single stage natural gas turbine, which is a typical method of providing peak power to electricity grids. The natural gas peaking costs are far higher than those anticipated when these systems were installed because the fuel costs have gone up from $2 per million Btu to almost $8 per million Btu (see Figure 1-6, Chapter 1).[22]

At least some solar technologies are on the threshold of an installed cost of $2 per peak watt at intermediate- and large-scales. At $2 per peak watt, the cost of solar electricity would be about 12 cents per kilowatt hour, well under peak power costs, and not much different than the cost of electricity generated using a natural gas combined cycle plant at a fuel cost of $8 per million Btu and delivered to the residential sector. The DOE's projection for 2015 of solar PV competitive with present-day large-scale commercial power plants comes in the context of rapidly declining solar PV costs and rapidly expanding global manufacturing capacity. As noted, the scale of manufacturing plants is also increasing, which is a key to cost reduction

The technological developments to make solar PV economical to supply peak and intermediate-level power have largely been accomplished with both thin film cells made of materials other than silicon as well as silicon cells using new manufacturing techniques or Fresnel lens concentrators. The issues remaining are increasing the scale of manufacture, and developing a wider infrastructure for manufacturing of the associated components, such as inverters, at larger scales. An analysis of the effect of very large-scale manufacturing of thin film technology – 2,000 to 3,500 MW per year of solar PV modules – commissioned by the National Renewable Energy Laboratory indicated that economies of scale could bring the overall cost, including installation, down to about $1 per peak watt for a 6,000 watt roof installation, including manufacturer's and retailer's margins. The largest portions of the cost reductions estimated by the authors were by analogy with cost reductions due to increase in manufacturing scale achieved in the flat panel display industry. One key ingredient was mass manufacture of the machines that make solar cells. One hundred lines of such machines were envisioned for a single plant.[23] One dollar per peak watt appears too optimistic for a residential rooftop system, given that costs of the parts, other than the solar cells, and of installation are unlikely to decline as much as the cells themselves. However, it appears reasonable that, with improvements in manufacturing technology, installed costs of $1 to $1.50 can be achieved in systems of 100 kW and larger. We have assumed $1.50 per peak watt in the reference scenario, which relies mainly on such systems.

The next few years will likely see which of the competing technologies will be manufactured at a large enough scale that the machines for the manufacturing can be mass produced. At that stage, one can expect that the costs of large-scale installations should be $1 to $1.25 per peak watt or so – yielding a power cost of about 6 to 7.5 cents per kilowatt hour. In short, the solar PV industry appears to be at about the same stage as wind was in the early 1990s, when it began to change from an industry with custom manufacturing of a few large-scale installations to a relatively mature industry today that can out-compete new nuclear power plants.

In the reference scenario for this study, we will assume that large-scale deployment of solar cells (on the scale seen for wind energy today) will not take place until about 2015 or 2020, though it may well do so before that. We assume an intermediate-scale installation cost of $1.50 (reflecting a mix of large-scale, intermediate-scale, and a smaller component of small-scale installations). Costs of storage and added costs for distribution are added as well (see Chapter 5 for details). As we will discuss, time-of-use pricing is an important policy tool for a transition to a renewable electricity system. It also best reflects market considerations in terms of cost of supply. A lack of time-of-use pricing is a reflection of improper market signals and the cause of significant market failures in the electricity sector.

We will incorporate all levels of solar electricity – very local residential (up to a few kW), medium-scale commercial (100 kW to a few MW), as well as central station (100 MW or more) – in our approach to a zero-CO_2 economy.

It turns out that a considerable part of the potential for solar electricity generation can be achieved on an intermediate-scale at the point of use – on rooftops, over parking lots, and if thin films get thin enough and cheap enough, simply by covering south-facing walls of buildings with photocells. We consider parking lot solar PV because of the potential scale of this resource and its many advantages in medium-scale applications. Let us first consider actual examples. Figure 3-4 shows a 235 kW installation for a 186 vehicle parking lot – or more than one kW per vehicle.

Figure 3-4: Kyocera "Solar Grove" – 25 Panels, 235 kW Total, 186 Vehicle Parking Lot.

Figure 3-5 shows a larger, 750 kW, U.S. Navy system in San Diego installed in 2002. It is easy to see that there is plenty of room to install additional solar energy capacity in that parking lot.

Figure 3-5: U.S. Navy 750 kW Parking Lot Solar PV Installation Near San Diego

Source: Courtesy of PowerLight Corporation

According to PowerLight, this installation is expected to avoid nearly a quarter of a million dollars per year of peak electricity costs:

> The 750 kW solar electric system was implemented as part of an Energy Savings Performance Contract (ESPC) project developed by NORESCO of Westborough, MA. The photovoltaic system was designed, manufactured and installed by PowerLight Corporation of Berkeley, CA. This photovoltaic system will produce approximately 1,244,000 kWh per year and is expected to save over $228,000 in annual operating costs by avoiding purchases of expensive peak electricity.[24]

Google is planning an even larger installation – 1.6 megawatts – sufficient to supply its headquarters with a large part of its electricity, in a combination of parking lot and rooftop deployment.[25]

Parking lot solar PV makes a great deal of sense for several reasons. Among them:

1. It does not require roof penetrations, reducing maintenance and the risk of leaks.
2. It does not require any new dedicated land.
3. It can be implemented on a scale that provides significant economies in installation costs.
4. It provides shade to parked vehicles, increasing comfort and reducing the need for air-conditioning at full blast when vehicles are started after being parked on bright summer days.
5. It increases the value of the parking lot.
6. Not least, grid connections in large parking lots (and rooftops) can be made compatible with vehicle-to-grid storage systems, discussed below. In these systems, parked electric vehicles or plug-in hybrids can supply power to the grid during peak daytime hours (for instance, on hot summer days), having been charged during off-peak hours at night. They could also be charged in the workplace during off-peak hours (for instance during night shifts or the early morning hours), with the same result. This also increases the value of the vehicles parked in the lot.

The land area devoted to parking spaces in the United States is very large. It has been estimated by the Earth Policy Institute at about 1.9 million hectares, or 19 billion square meters.[26] Most of these are not multi-story parking lots, but rather vast expanses of asphalt at shopping centers, offices, high schools, universities, airports, strip malls, supermarkets and other large stores, and the like, as well as private parking spaces. At 15 percent conversion efficiency, available today, parking lot PV installations could supply more than the electricity generated in the United States today. Of course, it may not be practical to use much of the parking area; some of it may be shaded much of the day, for instance. But parking lot solar PV installations could play a large role in a future electricity grid especially in the context of vehicle-to-grid (V) applications. Parked cars could exchange power with the grid, both serving as storage devices for times when excess electricity capacity is available and supply devices when the grid requires more electricity than the generation system can supply. Similarly, large flat commercial rooftops can also be used.

The first test of a V2G system is being started by Google and Pacific Gas & Electric (PG&E), the electric utility in the area, with a single Toyota Prius that has been converted by the addition of batteries and electronics to a plug-in hybrid. PG&E will control the charge on the batteries remotely, to test the system of charging the batteries when they are low and taking power from them when needed by the grid. [27]

Solar electric systems can also be used in more centralized installations. At 15 percent efficiency, a 1,000 MW plant in the Southwest (that is, in a favorable area for solar) would occupy about 300 acres, including 70 acres or so between solar PV arrays. Plant buildings and roads would be in addition to this area. Figure 3-6 (see color insert) is a map of the continental United States, published by the National Renewable Energy Laboratory, showing annual average incident solar radiation on a device that turns to face the sun. Figure 3-6 shows that there are large areas in the Southwest which are favorable to solar energy (more than 6 kWh per square meter per day). Much of the rest of the United States has an insolation rate of 4 to 5 kWh per square meter per day. The insolation values have been averaged day and night, over the entire year. The semi-arid and desert areas in the Southwest and West not only have the greatest incident energy, but also the greatest number of cloudless days. Those regions are therefore excellent candidates for central station solar PV, especially since this technology, unlike fossil fuel and nuclear plants, does not require cooling water. At 15 percent efficiency, a square meter of land with insolation at about seven kilowatt hours per square meter would generate about 400 kilowatt hours per year. Hence, an amount equal to about a trillion kilowatt hours – one-fourth of today's annual electricity output – could be produced on about 650,000 acres – a square with a side of just over 30 miles. With ancillary facilities, it would be a square with a side of about 35 miles.

Solar energy, of course, has in some measure a problem of intermittency, but in arid and semi-arid climates, this is not a significant issue, especially if solar PV is integrated with other energy sources. Solar insolation is much more predictable than wind on a hour-ahead, day-ahead, and seasonal basis. Moreover, it does not have the same kinds of micro-fluctuations that can create regulation problems on a time scale of seconds or minutes that wind energy does. Finally, being available in the daytime, it covers many of the peak hours, notably in the summer.

However, there are also certain periods of no sunshine when solar PV output is zero. Hence the problem of storage occurs on a diurnal time scale. Seasonal variations can also be considerable, the more so at higher latitudes. Figure 3-7 (see color insert) shows seasonal solar insolation variations, each value being averaged over a month (diurnal variations are taken into account in these averages). At 30° latitude (which runs through Texas, southern Louisiana, and northern Florida), solar insolation varies by a factor of almost two between the summer peak and winter trough. But in the United States the location of central station solar PV installations (or other solar installations) in the West and Southwest, two regions that combine availability of land and sunshine, would be feasible, since for most of the country the peak of demand occurs in the summer. Still, seasonal variation will be something of an issue since most of the land area of the United States is above 30° N.

Figure 3-8 (see color insert) shows the effect of nighttime lack of solar energy according to season for a zero-net-energy solar home in Virginia. The net effect of the seasons in balancing generation and demand on how much electricity is purchased and how much is fed back (exported) into the grid is quite complex. The June insolation daytime peak generation results in a high net feedback into the grid; but the export of electricity is about as high in October, when both demand and insolation are lower.

The graph "shows that even in the winter months a solar home is net exporting to the electric grid during the day and importing electricity from the electric grid during the early morning and evening hours. The time between 1300 and 1600 is the traditional peak for electricity particularly during summer months." (ERT 2005 page 11)

A part of the problem of diurnal and seasonal variation in solar energy can be dealt with by combining solar thermal power plants with heat storage as well as supplemental fuel use with solar thermal generation. Central station solar thermal plants use concentrators to focus heat on long pipes (parabolic troughs) or on a small area ("power towers"). There are nine power plants of the former design, between 14 MW and 80 MW, totaling 354 MW, operating in California that were installed between 1984 and 1990 by Luz International.[28] A variety of heat storage devices ranging from concrete and bricks to molten salt are being investigated, but none have been demonstrated in conjunction with a commercial solar thermal power plant. Capital costs for heat storage are estimated to vary between $30 for concrete and $130 per kilowatt hour-thermal for some phase-change materials. Since thermal energy must be converted to electricity with significant loss of energy, the capital costs of capacity to store enough heat to generate one kilowatt hour of electricity are significantly higher.[29] At $30 capital cost per kilowatt hour for concrete, assuming that the storage is used once every-day, the storage cost per kilowatt hour of electricity generated would be about 4 cents plus the cost of the solar thermal plant itself. In addition, there would be the operating and maintenance costs of the equipment associated with storing the heat – piping, pumps, etc.

Every energy source has its environmental costs, but when all is said and done, those associated with solar energy, even at a very large-scale of deployment, would be small. At present, the main environmental problems associated with solar energy arise from the emissions from fossil fuel plants that provide the energy to make the photovoltaic cells. Since crystalline silicon cells are the most energy intensive, the largest emissions, whether of heavy metals or CO_2 are associated with them.[30] They are higher than with wind energy due to the greater energy intensity of silicon cells.[31] Fresnel lens concentrators, which reduce the amount of silicon needed per unit of power generation, as well as newer techniques for manufacturing the thin strips of silicon needed for solar cells, will significantly bring down the energy cost of these cells. The emissions are lower with thin film cells mainly due to the lower energy manufacturing use per cell, despite their lower efficiency.

The indirect energy impact of solar PV, notably silicon cells, is declining due to more efficient use of silicon. Further, the indirect pollutant emissions are expected to be small once fossil fuels are eliminated from the energy supply. However, there will remain some impacts of mining, notably mining elements that are present in ores in small concentrations, as, for instance, with cadmium. Fthenakis and Kim estimate that these emissions would be quite small – 23.3 milligrams per million kilowatt hours – for cadmium telluride thin film PV, with the main impact coming from the production processes (production of the alloy and the PV cell itself) rather than mining. They estimate that mining impact is ~0.1 percent of the total cadmium emissions. The small mining impact is mainly due to the fact that the cadmium is a by-product of zinc manufacturing, with the main emissions being attributed therefore to zinc.[32] How such allocations might change in the face of very large-scale deployment of thin film solar PV must be evaluated. Recovery and reuse of the materials would greatly reduce their ultimate impact.[33] We note here that lithium-ion batteries, which would be used for electricity storage in V2G systems, can be recycled.

C. Biomass – Introduction

Solid biomass in the form of wood, crop residues, and cow dung still provides the bulk of residential fuel use for many or most people in developing countries, as it has for centuries. Biomass also provides the food for animals that still provide the main source of draft power for agriculture in much of Asia.[34] However, the use of biomass fuels directly in the form of liquids and gases on a large-scale has drawn considerable interest since the first energy crisis in the West in 1973, when OPEC increased oil prices and the Arab members of OPEC imposed an oil embargo on the United States, Western Europe, and Japan. The initial flurry of interest in the United States faded to some extent in the 1980s and then more so in the 1990s, with only a modest amount of ethanol derived from corn finding a niche in the automotive fuel market. A number of initiatives, including the possible use of the most productive plants, measured in terms of their efficiency of capture of solar energy, were abandoned. At least one country took a different path. Brazil persisted with ethanol production from sugarcane. Dual fuel cars are the norm in Brazil. Ethanol now supplies about 40 percent of motor vehicle fuel in Brazil.[35]

In the last few years, a number of factors, including rising petroleum prices and political and military turbulence in critical oil exporting areas, notably (but not only) in the Persian Gulf region, have caused a dramatic change in U.S. biofuel policy and production, centered on the production of ethanol from corn. President Bush featured ethanol production in his State of the Union speech two years in a row.[36] At the end of 2006, the ethanol production capacity in the United States was more than five billion gallons per year.[37] In his 2007 State of the Union speech, President Bush set a production target date for "renewable and alternative fuels," including ethanol, of 35 billion gallons for the year 2017.[38]

Biofuels can be a significant part of the energy supply. However, there are a number of fundamental issues that must be addressed not only to ensure long-term reliable and economical supply but also to verify that other serious problems, such as food insecurity, indirect large CO_2 emissions, or major economic inequities within countries or internationally do not arise as a result of fuel production from biomass. This is a very complex topic. The present report cannot do full justice to it. However, in view of the critical nature of the issue to energy supply, greenhouse gas emissions, land use, environmental protection, and other areas, it is important to consider it here to the extent needed in the context of an overall roadmap for a zero-CO_2 economy, including research and development, as well as infrastructural needs.

Basic considerations of the efficiency of photosynthetic solar energy capture under various circumstances are a good place to start. Solar insolation at mid-temperate latitudes at midday on a clear day provides energy at the rate of 1,000 watts per square meter.[39] The average over 24 hours is, of course, considerably lower due to a variety of factors, mainly no sunshine at night, considerably reduced insolation in the early morning and late afternoon hours, cloud cover, seasonal variations, and precipitation. As a result, the average annual insolation across most of the contiguous United States and Hawaii ranges from about four to about eight kilowatt hours per day per square meter.[40]

For food crops, the capture efficiency of solar energy is typically a fraction of one percent. For instance, corn yields are typically 8,000 to 10,000 kilograms per hectare[41] in the Midwest. The solar energy capture efficiency for a yield of 8,000 kilograms per hectare is about one-quarter of one percent.[42] Converting corn to ethanol results in about half or just under half of the energy value being in the ethanol; the rest is accounted for by co-products, like animal feed, and losses.

Low solar energy capture even at high food crop yields is only a part of the difficulty with the use of corn as a feedstock for ethanol production. A considerable amount of energy is needed to convert corn to ethanol – for instance, large amounts of steam are required. As a result of low solar energy capture, heavy use of fertilizers, and other inputs that are energy intensive, the net energy balance is not very good, even when the energy value of the co-products like animal feed is taken into account. A careful assessment of various studies on a commensurate basis indicates a range from approximately zero gain (energy used about equal to the energy output) to a net energy output of about 29,000 Btu per gallon (8 megajoules per liter). The latter is only 0.035 percent of the incident solar energy on the land. The energy input was estimated at 76,000 Btu per gallon (21.2 megajoules per liter).[43] Since coal, natural gas, oil, and electricity (largely derived from fossil fuels) are all needed for ethanol production from corn, and since other greenhouse gas emissions, such as nitrous oxide emissions due to nitrogen fertilizer use, also result from corn production, the greenhouse gas balance compared to gasoline is also rather poor. Some estimates of greenhouse gas emissions are actually higher than for gasoline, while others are somewhat lower. However, ethanol production does have a significant positive effect in reducing petroleum consumption, since much of the energy used in its production is in the form of natural gas, coal, and electricity.[44]

It is being rapidly recognized that the use of corn (and other food crops) for fuel on a large-scale can create serious competition with food. This already appears to be occurring as a result of the rapid growth of U.S. ethanol production. For instance a combination of demand for corn for ethanol in the United States production as well as local problems in market structure in Mexico has already contributed to a serious escalation in tortilla prices in Mexico:

> ...Although Mr. Calderón [President of Mexico] moved quickly, announcing a pact on Jan. 18 [2007] to freeze prices, the problem has not been resolved. Even with the pact, the news reports focused on the fact that the price ceiling for the tortillas of about 35 cents a pound was about 40 percent higher than the price three months earlier and contrasted that with the 4 percent increase in the minimum wage, which is still less than $5 a day.

> But because fewer than 10 percent of tortilla producers signed on to the agreement, the government had little power over those who did not. In some areas, prices have risen to 45 cents a pound. There is little more that Mr. Calderón can do to contain prices without huge expenditures for subsidies. Most analysts agree that the main cause of the increase has been a spike in corn prices in the United States, as the demand for corn to produce ethanol has jumped.

> But the uneven structure of Mexico's corn and tortilla industry here has also generated accusations – none of them proved – of hoarding and profiteering. Mexico's corn flour industry is controlled by just two companies, Grupo Maseca and Minsa. Under the pack, Grupo Maseca agreed to keep the prices for corn flour at 21 cents a pound. The government has promised to crack down on profiteers.[45]

In effect, a part of the burden of reducing oil imports by substituting corn-derived ethanol is being paid by the poor in Mexico. The global effects of rapidly increasing the use of corn, and possibly other food crops, such as cassava, which is a subsistence crop in much of Africa, for fuel ethanol could be devastating to the world's poor. Runge and Senauer have done a policy review of the issue going back to the 1970s and concluded as follows:

> The enormous volume of corn required by the ethanol industry is sending shock waves through the food system. (The United States accounts for some 40 percent of the world's total corn production and over half of all corn exports.) In March 2007, corn futures rose to over $4.38 a bushel, the highest level in ten years. Wheat and rice prices have also surged to decade highs, because even as those grains are increasingly being used as substitutes for corn, farmers are planting more acres with corn and fewer acres with other crops.

This might sound like nirvana to corn producers, but it is hardly that for consumers, especially in poor developing countries, who will be hit with a double shock if both food prices and oil prices stay high. The World Bank has estimated that in 2001, 2.7 billion people in the world were living on the equivalent of less than $2 a day; to them, even marginal increases in the cost of staple grains could be devastating. Filling the 25-gallon tank of an SUV with pure ethanol requires over 450 pounds of corn – which contains enough calories to feed one person for a year. By putting pressure on global supplies of edible crops, the surge in ethanol production will translate into higher prices for both processed and staple foods around the world. Biofuels have tied oil and food prices together in ways that could profoundly upset the relationships between food producers, consumers, and nations in the years ahead, with potentially devastating implications for both global poverty and food security.[46]

Runge and Senauer estimate that an additional 600 million people in developing countries could face malnutrition or starvation relative to trends in 2003, that is before the recent "biofuel mania."[47]

The integration of global markets and the rapid changes in production patterns and prices can result in serious problems in other areas as well. For instance, when the global balance of greenhouse gas emissions is taken into account, the use of food crops for fuel production can be much more damaging than revealed in an analysis focused at the national or regional level. One of the most dramatic examples in this arena is the increased emissions of carbon dioxide in Indonesia due to the export of palm oil to Europe for biodiesel production. When the per-acre yield of biodiesel alone is considered, palm oil appears to be one of the more attractive ways to produce biodiesel.[48] However, a recent detailed analysis shows that one metric ton of palm oil production on cleared and drained peatlands in Indonesia results in 10 to 30 metric tons of CO_2 emissions,[49] which is three to ten times more than the emissions from burning petroleum.

Ethanol from corn has provided two advantages so far in terms of guidance for policy. First, it has, after a considerable lull, re-focused attention on the potential large-scale use of biomass for fuel in the United States, which has the advantage of possessing a large, uncultivated land mass that is generally unsuitable for crops. Second, it has shown that an infrastructure for alternative fuels can be rapidly created, given the right policies. Of course those policies also need to focus on the appropriate technical, environmental, and economic choices. Producing fuel from food is already having deleterious effects and should not be encouraged by policy (see Chapter 7).[50]

D. Microalgae

Corn stover and other crop residues can provide inputs for ethanol production that would avoid some of the difficulties that are associated with the use of corn. However, large-scale production of liquid fuels from biomass or, for that matter, of solid fuels for electricity production would require a resource base that is considerably larger than that available from crop residues.[51] This restraint is strengthened when appropriate consideration is given to land conservation issues, which are important, among other things, for maintaining the soil's ability to continue to fulfill its role as a large reservoir of CO_2. Hence, while crop residues can and will likely play some role in the context of an economy with a large biofuels sector, they cannot play a central role in a large-scale biofuel supply. For the purposes of this investigation, we focus therefore on new biomass that is not associated with food crops.

There are two broad categories of biomass that could be cultivated for producing biofuels: grasses of various types and high productivity plants that grow in aquatic environments. As an example of the second type, microalgae exhibit prolific growth in a CO_2-rich environment. Microalgal productivity in such an environment in a sunny climate could be as high as 250 metric tons of dry mass per hectare per year, without using any artificial fertilizer other than exhaust from a power plant using fossil fuels.[52] Other plants that grow in nutrient rich environments, notably wastewater, at very high productivity in the range of 100 to 250 metric tons per hectare are duckweed and water hyacinth. The highest productivities are achieved in tropical or semi-tropical zones, though duckweed will also flourish for part of the year in the temperate zone. At the high end of productivity, the efficiency of solar energy capture of these plants is about 5 percent or about ten times that of the entire corn plant. It is about 20 times the efficiency relative to the solar energy capture in corn alone.

Demonstration-scale microalgae production using CO_2 from power plants has been carried out in two different contexts. The first used CO_2 from a 20 MW cogeneration plant at the Massachusetts Institute of Technology. The algae were not grown in open pools but rather in tubes slanted to face the sunlight.[53] The land area is minimized, the efficiency increased, and the quality of the algae is better controlled in this way. The algae apparently adapt to changing environmental conditions rapidly without a need for genetic engineering. The second is a small-scale bioreactor producing algae that has been operating in Arizona (Figure 3-9). A third demonstration plant has been installed at a coal-fired power plant in Louisiana (Figure 3-10).

Figure 3-9: Pilot Engineering-Scale Microalgae Plant at the Redhawk Gas-Fired Power Plant in Arizona

Source: Courtesy GreenFuel Technologies

Figure 3-10. Operating Demonstration Algae Bioreactor at a Coal-Fired Power Plant in Louisiana.

Source: Courtesy GreenFuel Technologies

It has been successfully tested using brackish and salt water. Isaac Berzin, who leads the research and development team for this technology for the company GreenFuel and also led the one for the MIT installation, has noted that the ability to use land of any quality and water of any quality are at least as important as the efficiency of solar energy capture. The target is a productivity of 100 metric tons per acre when the operation is commercialized (250 metric tons per hectare). The engineering-scale unit uses CO_2 from a combined cycle plant owned by Arizona Public Service, which is the largest electricity supplier in that state.[54]

A seven-day test at the MIT plant showed that daytime CO_2 removal was over 80 percent on sunny days and over 50 percent on cloudy and rainy days. Nitrogen oxide removal was in excess of 80 percent.[55] The engineering-scale unit in Arizona is on 0.3 acres of land. It operated in the spring and summer of 2007 in Arizona at the site of an Arizona power company's (APS) power plant. The expected breakeven price for a fully operational, large-scale plant is under $30 per barrel, without any subsidies or CO_2 credits.[56] Note that when the biomass is burned the CO_2 is released. Hence, microalgae, as a method of CO_2 capture from fossil fuel use, can result in large reductions in CO_2 emissions, but cannot by themselves result in a zero-CO_2 system. However, the same technology can also be used to capture CO_2 from electric power generating stations that use biomass as a fuel. Both uses of this technology are incorporated into the reference scenario (Chapter 5).

Since microalgae can be used to capture CO_2 from large-scale fossil fuel burning such as that in coal-fired and combined cycle power plants or cement plants and even combined heat and power systems, it can have medium-term impact in some major ways if it is sucessfully commercialized:

- Reduction of CO_2 (and NOx) emissions from existing fossil fuel power plants in the electric power sector.
- Reduction of industrial CO_2 emissions by CO_2 capture from cement plants, blast furnaces, and combined heat and power plants.
- Reduction of petroleum use (and hence oil imports) – in effect, CO_2 from coal, and natural gas combustion is combined with solar energy to produce petroleum substitutes. These substitutes could be various combinations of biodiesel and ethanol, depending on demand and the type of algae used.[57]

The very large capacity of coal-fired power plants, used to supply about half of the U.S. electricity, plus much smaller, but still important thermal uses of coal in cement and steel, are among the main reasons that the existing fossil fuel system has large economic inertia. In addition, natural gas use in central station power production, as a heat source in industry, and for combined heat and power production also results in considerable CO_2 emissions that could be captured in algae. The other very large sector of CO_2 emissions is, of course, the use of petroleum in transport, mainly land-transport, but also aircraft. While these emissions cannot be captured in biomass in any practical way, the fuel for them can be made from biomass, including algae production from the capture of power plant and industrial CO_2.

Algal bioreactors could capture most of the daytime emissions of CO_2 from large-scale sources. Nighttime emissions can only be captured if the CO_2 is stored and then passed through an additional bioreactor in the daytime. This necessitates local CO_2 storage in an underground reservoir. But the scale of the temporary sequestration is orders of magnitude lower than that required for long-term sequestration of CO_2, since storage capacity is needed for part of a day only, rather than for decades. Moreover, the risks that may arise from long-term storage are avoided.[58] The storage of nighttime CO_2 for daytime capture in algae would be akin to compressed air storage associated, say, with a wind farm, in which off-peak wind energy is stored at high pressure for generating electricity during peak and intermediate load hours. The technology of algae biomass production would likely first be commercialized for daytime capture, while the cost and technical issues associated with nighttime storage of CO_2 for daytime use are worked out. Overall, in sunny areas such as the Southwest, it may be possible to capture about 70 to 80 percent of the CO_2 in algae. The dry mass of algae is about double the captured mass of carbon, with the added weight being contributed by hydrogen, oxygen, nitrogen, and other elements.[59] With full implementation of CO_2 capture in algae, about seventy percent of the energy in coal could be captured in algae using bioreactors to convert CO_2, water, and other elements into biomass. [60] This can be converted into liquid biofuels, offsetting oil imports. The overall efficiency of liquid fuel production could be up to 10,000 gallons per acre per year.[61]

The carbon captured in the algae is emitted when the fuels are burned, for instance, in cars. The net effect is to reduce CO_2 emissions from the displaced petroleum consumption. Conversion of microalgae to liquid fuels at acceptable cost at or near the targeted efficiencies remains to be demonstrated. A commercial plant has not yet been built.

In the longer term, as fossil fuels are phased out, the approach of using CO_2 from fossil fuel combustion for algae production is not compatible with a zero-CO_2 economy, since the CO_2 will eventually be emitted from vehicles or other machinery. However, microalgae can also grow in saline, nutrient rich waters, such as run off flowing into the Salton Sea, as well as in ponds. In the long-term, transportation will be supplied by (i) electricity, (ii) hydrogen produced from wind or solar energy, or (iii) biofuels. Fuel can also be produced from landfill methane, forest wastes, food wastes and other similar sources of biomass.

E. Grasses

Switchgrass, a high-yield, perennial prairie grass that can be grown in a variety of circumstances, has been investigated recently as a prime candidate for anchoring the supply of biofuels to overcome the limitations of ethanol from corn. A seminal report was issued by the Natural Resources Defense Council in 2004, which estimated that by 2050 the United States could be producing 7.9 million barrels a day of biofuels (in petroleum equivalent) using this approach.[62] The report cautions that switchgrass is one good candidate for creating such a supply but that further work is needed. Switchgrass has some ancillary environmental advantages:

Switchgrass also offers low nitrogen runoff, very low erosion, and increased soil carbon–which is actually enhanced when the crop is harvested. Switchgrass also provides good wildlife habitat. It is likely that such benefits are not limited to switchgrass, although other crops were not investigated in any detail.[63]

The current productivity of switchgrass is estimated to be about 10 to 12 metric tons per hectare per year over a variety of growing regions and that by 2050 this could be about 25 to 30 metric tons per hectare per year by crop selection done without genetic engineering.[64] Farrell et al. have estimated that if current approaches to converting cellulosic material to liquid fuels can be made economical, that the energy and greenhouse balance of switchgrass would be very favorable.[65] The ratio of output energy to input energy is estimated at 8.2 and the emissions of greenhouse gases are estimated at 11 grams carbon equivalent per megajoule compared to 94 for gasoline.[66] Growing fuel crops on marginal lands is also possible and, done appropriately, it can provide measurable increase in carbon sequestration in the soil, without the use of expensive and energy intensive inputs such as fertilizers and pesticides. This approach would avoid the use of high quality land and inputs for biofuel production while providing larger collateral environmental benefits.[67] The cultivation and harvesting of biomass in such a way as to sequester carbon in the soil in measurable ways is a crucial part of the process of developing the large-scale use of cultivated biomass in the energy system. It is also important for other types of biomass in case net removal of CO_2, beyond zero-CO_2 emissions, is pursued.

The land requirements implicit in using grasses at productivities of 25 to 30 metric tons per hectare (10 to 12 metric tons per acre) as the mainstay for biofuel production would cause significant, possibly unacceptable, land use impacts (see Chapter 5). It is, therefore, crucial to tap into higher productivity biomass, including, but not only microalgae, to produce liquid fuels and industrial feedstocks. Alternatively, direct production of hydrogen from solar energy could replace a large portion of the biofuel requirements with much smaller land requirements, provided the methods can be made economical (Chapter 6).

The initial stage of development of the technology of the use of solid biomass as fuel is occurring in the context of co-firing biomass with coal. This can be done for power production only or for combined power and liquid fuel production. Co-firing in IGCC plants with coal and biomass has already been tested, for instance, by Tampa Electric. The flow diagram of the plant is shown in Figure 3-11.

Figure 3-11: Flow Diagram for the Tampa Electric Test of Co-firing Biomass with Coal and Petroleum Coke

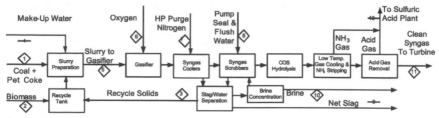

Source: Tampa Electric, 2002, page 3. Reprinted with permission of Tampa Electric Company.

The proportion of biomass burned in the Tampa Electric test was very small – only one percent. It was to test whether there was an increase in sulfur dioxide or NOx emissions from the power plant due to an introduction of biomass feed. The proportion of biomass was kept high enough for the measurements of the pollutants of concern to be statistically significant.

It is important that IGCC technology, that can use mixtures of biomass and coal and that can run on biomass alone to produce power and liquid fuels, be developed. In the recommended scenarios in this report, we do not assume the use of coal. However, it is important to note that the requirement for liquid and gaseous fuels in transport and industry is likely to remain very large. Hydrogen produced from renewable electricity can be used in transportation, in whole or in part. However, portions of such use, notably for aircraft, require long-term development.

F.   Other High Productivity Biomass

Even with substantial hydrogen and direct electricity use in transportation, there is still likely to be a large requirement for liquid and gaseous fuels for transport and industry in a zero-CO_2 economy. It is important to plan for about 15 to 20 quadrillion Btu per year of such fuels, even in an economy where efficiency increases result in a steady absolute decline in energy use (See Chapter 6).

Production of large amounts of biofuels using mainly switchgrass or other prairie grasses would likely create unacceptably high land requirements. It is important, therefore, to consider whether there are other sources of high productivity biomass, comparable to microalgae, which do not require an input of high CO_2 gases. The water hyacinth in semi-tropical (and tropical) climates is one such plant. (See Figure 3-12 in color insert.) Duckweed is another. The latter also grows well in temperate climates. Both of them grow prolifically in wastewater rich in nutrients. The productivity of water hyacinths in semi-tropical climates, if they are harvested regularly, is comparable to microalgae grown in tubes with CO_2-rich exhaust from power plants – that is, about 100 dry metric tons per acre. Indeed, at up to 17.5 wet tons per hectare per day, it may be the most productive plant on earth.[68] Under the right conditions, water hyacinths can produce as much dry matter in two weeks as switchgrass produces in one year. The high productivity depends on water that is rich in nutrients – nitrogen and phosphorous. These nutrients are, of course, present as pollutants in wastewater treatment plants and in run-off from agricultural lands.

In terms of the efficiency of solar energy capture, water hyacinths can achieve efficiencies up to 5 percent, which is several times the total biomass efficiency of most crops (which need energy inputs and artificial fertilizers) and two times or more than the biomass output of sugarcane.[69] In point of fact, without plant breeding or other intensive research to increase productivity, the efficiency of solar energy capture of water hyacinths is only about a factor of three lower than that of today's commercial solar PV cells. It is ten times higher than the entire corn plant.

In practice, the prolific productivity of water hyacinths has caused it to be regarded as a nuisance weed or worse, and for good reasons. It can choke waterways, requiring large expenditures for periodic removal. Mosquitoes may breed in infested waterways more easily, with attendant health risks. Further, the plants are killed by sustained temperatures (for about 12 to 24 hours) below about 24°F.[70] However, the ability of water hyacinths to soak up nutrients has also been seen as a potential boon in wastewater treatment and in treatment of natural ecosystems that have become seriously damaged by eutrophication due to nutrients in agricultural runoff. Hence, so far, experimental and demonstration projects with water hyacinths have centered on their effectiveness in wastewater treatment, both public and industrial, rather than as an energy source.

In the 1970s, the National Aeronautics and Space Administration initiated a project in Bay St. Louis, Mississippi, to try to address a problem of heavy metals in wastewater discharge for its National Space Technologies Laboratories (NSTL). Conventional treatment did not result in consistent compliance with EPA standards.[71] A lagoon of just over half-an-acre was constructed to receive and treat about 25,000 gallons per day of water, with a retention time of 20 days. Even with only chemical wastes from photography laboratories in the discharge water, the water hyacinths grew rapidly – by about five-fold, from an initial 20 percent stocking, in four weeks. Silver was the main metal pollutant in the effluent water. The results are worth quoting at length, not only because of the potential for wastewater treatment and energy, but for reducing heavy metals pollution and, indeed, their possible recovery and recycling.

> The water hyacinths proved to be a very effective filtration system for cleaning wastewater containing a complex chemical mixture. Organics, heavy metals and other elements were effectively removed from the wastewater by plant root sorption, concentration and/or metabolic breakdown... Trace elements entering the lagoon system were effectively removed to levels which comply with PHS [Public Health Service] recommendations.

Even the hardy water hyacinth is not immune to heavy metal pollutants. Approximately every eight weeks during the summer, the leaf tips began to turn brown and curl, indicating that the plants had sustained permanent metabolic injury from the environmental pollutants....

Since the plant stems and leaves, as well as its roots, were found to contain heavy metals, no part of the harvested plants can be used as feed or fertilizer. However, the harvested plants can be used safely for the production of biogas. Whole harvested plants (or remaining sludge, if biogas is produced) should be put in a pit especially designed to eliminate ground water infiltration. Such a pit is planned to be utilized at the NSTL zig-zag lagoon. Over a period of years, the heavy metals in the pit may accumulate to levels high enough that their extraction becomes economically feasible. Such small "mining" operations – particularly of silver – may prove to be an efficient method of recycling valuable metals for industrial use.[72]

There have been a number of demonstration projects using water hyacinths for public wastewater treatment.[73] Most of these were in small to medium systems where the biomass product was a liability, since it had to be composted or otherwise disposed of. Mosquito control was achieved partially through stocking of mosquito fish or completely through aeration, which also eliminates odors and allows high nutrient loading of the influent water. In colder climates, other very high productivity plants like duckweed and cattails have also been used. A mix of plants, using cold-resistant plants in the winter and water hyacinths in warmer seasons can also be used.

Experiments to produce biogas using water hyacinths have been conducted by NASA and others. The NASA research indicates that a mixture of plants, for instance, water hyacinths and duckweed, would produce better results, than either alone.[74] Using plants like duckweed may also be desirable in some areas for other reasons. Water hyacinths do not grow in brackish water, but other plants, such as duckweed, do.

The amount of effort into actually demonstrating the use of high productivity plants has been minuscule – so tiny that it is not on the radar screen of energy policy. Yet, their basic biological and solar energy capture properties indicate that they have the potential to:

- Greatly reduce the land area needed to grow biomass,
- Combine water treatment with very efficient biomass production for use in IGCC systems to produce electricity, hydrogen, or liquid fuels,
- Combine biomass production of various kinds by using water hyacinths, duckweed, etc., in IGCC systems, with the CO_2 effluent being used to cultivate microalgae for liquid fuel production – probably the most efficient combination,
- Provide a source of animal feed, if grown in wastewater that is free of heavy metals,[75]
- Provide the possibility of CO_2 capture from the atmosphere and sequestration of a solid material rather than CO_2 gas, in case negative CO_2 emissions policies are required in the future, and

- Provide the potential in industrial and urban wastewater treatment systems of recovering heavy metals for reuse in the economy.

The above list is not presented with the idea that this is some kind of a silver bullet, but to indicate the possible potential of an area that has received almost no attention in energy policy. When properly situated, aquatic plants could, in combination with other approaches, provide a significant portion of the energy supply in environmentally sound ways. Figures 3-13 and 3-14 show the areas where two of the candidate plants can be grown and the length of the growing season.

Figure 3-13: Areas Suitable for Water Hyacinths Systems

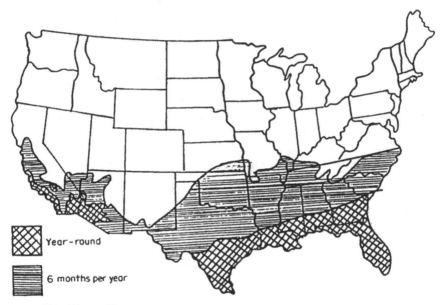

Source: EPA 1988 page 50

Figure 3-14: Areas Suitable for Duckweed Systems

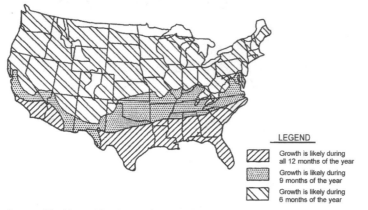

LEGEND

Growth is likely during
all 12 months of the year

Growth is likely during
9 months of the year

Growth is likely during
6 months of the year

Source: Zirschky and Reed 1988. Copyright © 1988 Water Environment Federation: Alexandria, Virginia. Adapted with permission.

The approach needs to be implemented with the sophistication that is possible with large-scale application and with the specific aim of optimizing the various outputs that are possible. The optimization will be different in different areas of the country. In some areas, land use and climatic factors may make the approach unsuitable locally. At the same time, if compressing and piping CO_2 for a couple of hundred miles is seen as feasible or even necessary for climate protection policy, it is even more worthwhile to explore the piping of wastewater to warm areas to produce clean water and achieve high efficiency solar energy capture in biomass.

G. Some Conclusions about Biomass

Even the above brief survey demonstrates the complexities of biofuel production as well as its immense potential. Some principal conclusions are, however, clear in the context of this report examining the feasibility of a zero-CO_2 economy:

- Food crop based approaches to biofuels requiring heavy inputs are not suitable for large-scale biofuel production if the main aim is to reduce greenhouse gas emissions. They are also not a very good choice due to low net energy output. Moreover, they can and often do create other social and environmental damage that is difficult or impossible to remedy. We will not consider food crops as a source of biofuels for the United States in this book.
- Cellulosic biomass from crop residues may provide a modest fraction of U.S. biofuel requirements, with appropriate cautions, but is unlikely to be a major source, defined as a few million barrels of petroleum equivalent a day, or more.
- Microalgae, used to capture CO_2 from fossil fuel power plants, could make a significant contribution to liquid fuel supply. Microalgae can also be grown in CO_2 captured from solid biomass burning as fossil fuels are phased out.

This technology needs full-scale demonstration. Storage of nighttime CO_2 and additional production of microalgae in the daytime can also be accomplished.

- Grasses can be cultivated on marginal lands in a manner that would not put fuel in competition with food. Their productivity is lower than microalgae, but they have the merit of capturing atmospheric carbon dioxide and can therefore be used in the long-term as part of a negative CO_2 emissions scheme. That is, combustion of biomass can, in principle, be accompanied by CO_2 capture and sequestration. They do not need a special source of CO_2; with appropriate crop selection and rotation, inputs such as fertilizers can be avoided or minimized.
- Aquatic biomass varieties grown in nutrient-rich wastewater, such as water hyacinths and duckweed, have enormous potential due to their high yields (comparable to microalgae). The technologies have been tried but their application for energy production potential has not been demonstrated on a significant scale.

H. Solar Hydrogen

There are many ways to produce hydrogen from solar energy. Many of them involve production of some kind of feedstock, such as glucose or some form of biomass, produced using solar energy. The feedstock is then processed, in some cases with the use of solar energy, to produce hydrogen. Biomass, such as aquatic plants and microalgae, can also be converted into carbon monoxide and hydrogen in a gasification plant similar to those being used in the Integrated Gasification Combined Cycle technology that has been developed for coal. These can then be turned into CO_2, water, and hydrogen, after which the hydrogen is separated from the other gases. Overall, this method is a special application of biomass production for energy.

Hydrogen is produced commercially today for industrial applications from natural gas, of which methane is the principal component. Hence, the same can also be done using landfill gas, which also has methane as its principal constituent (though in lower concentrations than natural gas). However, this would remain a relatively small source of hydrogen, since the source material is not very plentiful relative to energy requirements.

Direct solar hydrogen production methods include:
- Biological hydrogen production, using algae (photolytic hydrogen production)
- Photoelectrochemical hydrogen production – where various inorganic materials are arranged into solar cell type of devices, but instead of producing electricity, they split water into hydrogen and oxygen
- High temperature, solar-energy-driven systems that split water into hydrogen and oxygen, using catalysts.

For the most part, using solar energy to produce hydrogen directly is still in the laboratory stage of study. For photolytic hydrogen production using algae (see Figure 3-15 in color inserts), high efficiencies have been achieved in turning incident light energy into chemical energy, but the hydrogen production rate is still low, making for low overall efficiency. Higher efficiencies have been achieved with photoelectrochemical production and high-temperature catalytic splitting of water.[76]

To compete with gasoline at $3 per gallon, the delivered cost of hydrogen should be about $3 per kilogram (since one kilogram of hydrogen is approximately equivalent in energy terms to a gallon of gasoline). Of the approaches mentioned here, the IGCC approach is perhaps closest to commercialization, since most of the technological development has already been completed. However, the economics of the process will depend in part on the efficiency with which the feedstock biomass captures solar energy. This is the principal determinant of land requirements. Biomass, such as prairie grasses, could be used on a modest scale to produce hydrogen, but the land use implications of growing prairie grasses would not be qualitatively different than producing liquid fuels. Significant work remains to be done in regard to technology development before reliable cost estimates can be made.

The Department of Energy's target efficiency for photoelectrochemical hydrogen for 2010 is 8 percent – that is, the energy content of the hydrogen would have eight percent of the energy content of the incident solar energy.[77] This is very high efficiency – higher than that of any type of biomass. Further, unlike solid biomass, hydrogen can be used directly in internal combustion engines. High temperatures generated by solar concentrators can also be used to produce hydrogen and show promise of high efficiency. The DOE's target for the year 2015 is a cost of $3 per kilogram, which would be competitive with gasoline at current prices (July 2007).[78]

Direct hydrogen production methods, notably the photoelectrochemical and high temperature splitting of water have the potential to greatly reduce land requirements for a renewable energy economy relative to the reference scenario. This is one reason that one or both of these methods, and possibly others that can have comparable efficiencies of hydrogen production (five percent or more) can provide the basis for a partial hydrogen economy.

A mixture of biofuels produced with high efficiency and direct solar hydrogen production, with as large a component of the latter as possible, would be a preferred way of achieving a renewable energy future with the low environmental impacts relative to other biofuel scenarios. This is because the composition of most liquid biofuels is similar to that of petroleum-based fuels in that they consist of hydrocarbons. Burning them therefore would still raise pollution issues of unburned hydrocarbons, carbon monoxide, nitrogen oxides, and in some cases particulates as well. While all of these have been and can be further reduced, the use of hydrogen completely eliminates all but some nitrogen oxide emissions. Further, direct solar hydrogen production does not involve such air emissions in its manufacture. Therefore, in terms of urban air quality and the reduction of emissions from industry, hydrogen made directly from solar energy is a preferred energy source and should be developed.

I. Wave Energy

While the potential for generating electricity from the motion of waves is nowhere near as large as that of wind or solar energy, it could be an important source in some coastal areas. In contrast to offshore wind, which has faced considerable opposition in some areas, such as Cape Cod, Massachusetts, due to the high visual profile of the towers, the profile of wave generators is very low – they float on the surface of the water. Another advantage is that wave energy is more steady and forecastable, so that there is less of an issue with intermittency than there is with wind energy.

A study by the Electric Power Research Institute concluded that Hawaii, Oregon, northern California, and Massachusetts would likely be the first areas that could achieve economics on a par with wind energy. In contrast to the latter, wave energy is still in the early stages of large-scale demonstration. The potential is considered to be in the tens of thousands of megawatts.[79] In this study we assume that it will be included under the rubric of "geothermal and other" energy supply estimates in the future.

J. Hot Rock Geothermal Energy[80]

After the 1973 energy crisis, many energy research projects were initiated at the national laboratories, besides the establishment of a dedicated laboratory, now known as the National Renewable Energy Laboratory.[81] One of the most important projects, potentially, was one to investigate the feasibility of tapping the heat in high temperature rocks in some geologic formations for generating electricity. The project was carried out at Los Alamos National Laboratory for about two decades but closed down in the 1990s.

Two great advantages of hot rock geothermal technology (known more formally as "Extended Geothermal Systems" (EGS)) are (i) that it can provide baseload power, and thus be a critical part of reducing reserve or storage requirements in a system with intermittent sources and (ii) that it is far more widely available than conventional hydrothermal geothermal energy. The latter consists of water that is heated deep in the earth that can be brought to the surface and flashed into steam to drive a turbine. It is an important regional resource, for instance, in California. But it is far more limited than the heat in rocks at depths of 3 to 5 kilometers (10,000 to 16,500 feet). If this heat can be tapped for power production, geothermal energy could become a much greater energy resource. The main idea behind hot rock geothermal energy is to inject fluids into fractures in the hot rock zone and then pump the heated fluids back to the surface where they are then used to generate electricity in a manner similar to the way hot geothermal water is used today.

Deep drilling technology, developed among other things for oil and gas production, can be used in producing hot rock geothermal energy. However, much research, development, and demonstration remains to be done in every area from drilling to reservoir management to power production. An expert panel reviewing the technology has recently (2006) concluded that

> A cumulative capacity of more than 100,000 MWe from EGS can be achieved in the United States within 50 years with a modest, multiyear federal investment for RD&D in several field projects in the United States.[82]

For reference, 100,000 MW is approximately equal to the installed capacity of nuclear power plants today. This is an especially significant amount of power in any context, including that of the present study due to its ability to provide baseload generation. IEER's reference scenario assumes that about one-fourth of this amount will be developed as baseload capacity by 2050, with the first large plants coming on line in the 2020-2030 decade.

K. Energy Storage Technologies

Given the large part that wind and solar energy will play in a renewable energy economy, storage technologies will be critical to the reliable functioning of the electricity system. At present, with low penetration of these two sources, no storage is necessary since reserve capacity can be supplied in other ways. For instance, as we have noted, the excess capacity of natural gas-fired power plants can serve as a standby for wind, and it can also serve the same purpose for central station solar power plants. The Luz International central station solar thermal power plants have the capacity to burn natural gas at night to supply around the clock energy.[83] A new installation of that type would likely not need such a capability. It would probably be cheaper to have a contractual arrangement with an existing natural gas fueled combined cycle power plant operator to provide the needed energy in the evening hours.

However, in the final analysis, natural gas cannot continue to serve this function (except as a contingency) if fossil fuels are to be phased out (leaving aside, for the moment, the potential for CO_2 sequestration). We have already mentioned the possibility of heat storage in various media, such as concrete, in the context of central station solar thermal generation. We will not discuss it further here. Rather, we take up three other storage technologies:

1. Batteries
2. Capacitors
3. Compressed air

In addition to these sources, we assume that existing reservoirs and hydropower stations can be managed to complement wind energy by limiting their use to periods when the wind is not blowing but the electricity demand is still present.[84] We recognize that there are considerations other than electricity generation in the management of dams and reservoirs, such as irrigation, flood control, or endangered species protection. Combining solar, wind, hydropower, and combined cycle natural gas-fired power plants into a single system that is optimized could provide the added flexibility that is needed for multiple uses of water in the reservoirs. With a combination of sources, existing reservoirs can also be used for pumped storage. Some storage issues are discussed in Chapter 5 in connection with demand-side management in the electricity sector.

1. Batteries

Storage of electricity in batteries has been traditionally associated with lead-acid batteries, which are inefficient and heavy, but which have long had the merit of being cheap compared to other batteries. Lead-acid batteries are used, among other things for Uninterruptible Power Supply (UPS) in applications where even small discontinuities in energy supply for a few seconds can be very expensive. Batteries can supply a large amount of power for short periods of time (a virtue that has made them ubiquitous for starting cars). But they are not durable enough to be charged and discharged repeatedly, which is a requirement for electricity storage in a renewable electricity system.

In recent years, a number of new candidates have come into the market, such as nickel-metal-hydride (NiMH) batteries that are used in hybrid cars (such as the Toyota Prius). But these, too, have a very limited storage capacity; moreover, they are expensive. The most promising candidates for large-scale energy storage are new designs of lithium-ion batteries. These are similar to other batteries used in cell phones and many other portable devices. The new varieties do not use carbon, a source of safety concerns (and a reason for the recalls of lithium-ion batteries used in many laptop computers). Lithium-ion batteries with lithium-iron oxide and lithium-titanium oxide electrodes have a number of properties that make them suitable for all-electric cars as well as plug-in hybrids:

1. High storage capacity per unit weight – at present about 100 to 120 watt-hours per kilogram and expected to go up to about 180 Wh/kg (about six times the energy density of a lead-acid battery)
2. Capacity to be charged and discharged 10,000 to 15,000 times without significant loss of performance (applicable to the Altairnano battery)
3. High efficiency of charging and discharging
4. Ability to withstand deep discharge repeatedly
5. Satisfactory performance on safety tests (Altairnano battery)
6. Ability to be charged in a relatively short period of time (10 to 15 minutes) with appropriate heavy-current equipment.[85]

Such batteries have the kind of performance that could make all-electric cars economical in the next decade. The main requirement is that the cost needs to be brought down by about a factor of 5 from the present $1,000 per kilowatt hour of storage to about $200 per kilowatt hour. At the former cost, a car with a range of 200 miles would incur $40,000 in battery cost alone. However, these are still more or less custom-made batteries that do not have high volume manufacturing. The processes to make them are new. It is anticipated that with the kind of process improvements that are normal in manufacturing for a maturing technology and with a large enough scale (tens or hundreds of thousands of cars per year), such a cost reduction should be achievable in the next decade.[86]

The possibility of using passenger and commercial vehicles to exchange power with the electricity grid, and hence for vehicles to serve as an energy storage medium, was first analyzed in a 1997 paper by Kempton and Letendre,[87] according to a University of Delaware research project.[88] Passenger vehicles are usually parked. They are used a very small proportion of the time – typically 5 to 7 percent – creating the possibility of a vehicle-to-grid (V2G) system. Further, utilities could also contract with corporate and governmental owners of fleets of vehicles. These institutions have reliable ways to estimate the patterns of usage of their vehicles, which can then be partly matched to the requirements of a utility.

The installed power of engines in cars and light trucks is well over an order of magnitude more than that of the entire U.S. electric power system. Therefore, only a small fraction of vehicles is needed for energy storage for a vehicle-to-grid system to function reliably. For instance, at 10 kW per vehicle, 10 million vehicles would supply a standby capacity of 100,000 megawatts, the equivalent of 100 large nuclear power plants. Yet, 10 million vehicles would be only about three percent of the total number of vehicles projected for 2050. With fully or partly electric vehicles, a V2G system could store energy during off-peak hours and supply it during peak hours.[89] Or it could supply standby capacity for wind-generation to compensate for its intermittency. As discussed in Chapter 2, the marginal cost, and the implicit CO_2 price, of such a system could be low, if the vehicles themselves are economical.

There are, of course, a number of issues associated with the development and reliable functioning of a V2G system:

1. Will the energy stored in mobile vehicle batteries be available to the grid when it is needed?
2. Is the electricity distribution system robust enough to handle the amount of power that would run through it in a system with a high proportion of intermittent renewable sources?
3. How will the Independent System Operator, who must ensure that the stability of the electricity grid and the demand and supply are matched, communicate with vehicles when they are plugged into the grid and manage the system to ensure the right amount of power exchange to keep the grid functioning at all times?
4. Will the batteries last?
5. How could vehicle users be assured of sufficient charge remaining in their vehicles to be able to use them when they are needed?
6. What about rush hour?

These are critical questions and the feasibility of V2G systems depends on the answers. Yet they are not as daunting as they seem at first. For instance, the kind of satellite communications that have made global positioning systems (GPS) cheap and reliable enough to be available in individual cars can also be used for communicating with vehicles. Cell phone towers could also be used. High frequency signals sent through the electricity grid are also a possibility.

So far as the distribution system is concerned, it may be impractical, at least in the initial stages, to use individual homes as hookup points for V2G systems, though this may not apply to certain kinds of residential developments. For instance, a development in Atlanta was created as a community, with open spaces, a large, leased vegetable plot where locally grown produce is supplied on a commercial basis to residents, etc. One feature of this development is that there are only walking lanes in the community and a parking lot at its entrance. This feature of the community was not created for energy purposes but to make the spaces in the community safe for children and free of cars. But with dozens of vehicles parking in a single area, it would be much more practical to consider installing an infrastructure for exchanging power with the grid or even just for quick charging of plug-in hybrids and all-electric vehicles.

As noted in the section on solar power above, one principal hub of a V2G system could be the parking-lot/rooftop solar system that has V2G infrastructure installed with it. The two can be developed independently, as well. The number of vehicles in such situations could be estimated relatively easily. This is a scale where the installation of the communication with the Independent System Operator could be economical. With a diversity factor between various building and parking lots across a region, planning of power system resources should be possible in a reliable manner. In other words, with sufficient parties participating, the minimum number of V2G vehicles plugged into the grid at any time can be computed with a higher degree of confidence.

As noted above, the first test of a V2G system will be carried out in a collaboration between Google, whose Silicon Valley headquarters has rooftop and parking lot solar PV, and PG&E, the electric utility that serves the area. Google has purchased a plug-in hybrid (a converted Toyota Prius) whose batteries will be controlled by PG&E when it is parked.[90]

The costs of the infrastructure, apart from the batteries, have been estimated for a 5,000 car system at about 0.5 cents per kilowatt hour.[91] There are different estimates of losses for a charging and discharging cycle, which the utility would experience. Tesla Motors cites a value of 86 percent for its battery pack, while Solion, which makes battery systems for racing cars, has stated that the charge discharge efficiency of a single cell is 99 percent. We will assume a 90 percent efficiency for a practical charge-discharge cycle in the year 2050.[92] Since the batteries would be charged off-peak, the cost of electricity losses is on the order of 0.5 cents per kWh or less (with an off-peak electricity cost of five cents per kWh or less). The overall cost of the V2G system would therefore be expected to be one cent or less per kWh plus the payment to the owner of the battery and the parking spot. Overall, the cost of V2G storage of electricity and re-supply to the grid at peak and intermediate load times would be expected to be a little over one cent per kWh if there is sufficient competition to supply the V2G service.

The cost of a V2G system with batteries would be quite large unless the batteries can withstand charging and discharging, without significant deterioration in performance, in excess of the number of times that such charging would be needed for the use of the vehicles themselves. For instance, if a car is charged every 100 miles, the annual number of charges would be typically 150 to 200, which gives a total of 1,500 to 2,000 charges over an expected ten-year life of the vehicle. Typical batteries today can withstand charging a few hundred to 1,000 or 1,500 times. With such batteries, a V2G system would impose battery depreciation costs, which would markedly affect the viability of the system. One reason that V2G has been considered to be feasible in this study is that newly designed lithium-ion batteries now being installed in vehicles have been successfully tested for their ability to endure over 10,000 charging cycles. For instance, the lithium-ion battery with a lithium-titanium oxide electrode manufactured by Altairnano in 2006 has been tested over 15,000 deep discharges with 85 percent capacity still remaining after the tests.[93] This is 15 to 20 times the number of times a typical battery can be discharged and recharged. With such performance, the marginal battery cost imposed by a V2G system is close to zero (though, of course, the owner of the battery would reasonably want compensation for the service provided to the grid). The batteries are being installed in all-electric pickup trucks made by Phoenix Motorcars in 2007.

Lithium-ion batteries, which can be recycled, have also begun to be used in custom conversions of hybrid cars into plug-in hybrids. Hybrid cars use batteries to store energy recovered from braking and deceleration. The batteries store sufficient energy to enable a car to run on electricity only for short distances.[94] The addition of batteries can extend the electricity-only range, which reduces the use of gasoline, increases overall efficiency (since the electric part of the car is more efficient than the gasoline part), and reduces CO_2 emissions. Google's plug-in hybrids have been instrumented for measuring the gasoline and electricity consumption. As of July 8, 2007, the average mileage per gallon of gasoline was 73.6; in addition, the cars used about 0.12 kWh per mile of electricity. Plug-in hybrids using lithium-ion batteries could provide an opportunity for widespread demonstration of V2G technology in the next five years, if governments and corporations decide to purchase them in large enough numbers. Major automobile manufacturers have expressed various levels of interest in plug-in hybrids; some have announced specific models that will be made, but none have announced plans for large-scale production.[95]

2. Capacitors

Like batteries, capacitors store electricity, but they do so differently. Batteries store charge chemically, while capacitors store electrical energy by storing an electric charge on electrodes separated by an insulating material. As with a battery, there is a voltage difference between the electrodes, and the stored energy can be recovered by discharging the capacitor through a load, like an electric light or an electronic circuit. The amount of energy stored is proportional to the square of the voltage difference and the area of the electrodes that store the charge.

Capacitors have some very distinct advantages and disadvantages as energy storage devices. They are very efficient (95 plus percent efficiency is possible) and hence expensive electricity is not wasted in charging and discharging the device. They are also the fastest devices. A capacitor can be charged and discharged in seconds or fractions of a second. Batteries take a long time to charge and even with the most recent advances in lithium-ion batteries, the charging is anticipated to be 10 to 15 minutes with special equipment and several hours when plugged into a residential outlet.

There are a number of reasons why capacitors have not become central features of renewable energy systems. The energy density of even the best capacitors, known as ultracapacitors (or ultracaps, for short) is only 4 to 6 watt-hours per kilogram, compared to five to seven times as much for a lead-acid battery and 30 times as much for a lithium-ion battery. They also use expensive materials. The combination, of course, makes ultracaps bulky and expensive, and therefore unsuitable as the main energy storage device in vehicles. However, the speed of charging and discharging enables such devices to be used where the quality of power is at a premium and space is not – for instance, as voltage stabilizers at times of peak power demand.[96]

Ultracaps can also serve a useful role in electric vehicles and plug-in hybrids. A small ultracap storage capacity can serve the function of storing the energy recovered during regenerative braking and provide the energy for quick starting from a stop. A combination of small capacitor storage and a main battery storage system may make for more durable electric vehicles and better performance; it is in the initial stages of commercial exploration today. One company, AFS Trinity, has announced that it will manufacture an "extreme hybrid" which is a plug-in hybrid that uses a combination of a gasoline engine, batteries, ultracapacitors, and a flywheel to optimize the operation of the car for getting better performance from the batteries and the entire electrical portion of the vehicle.[97] Where weight is not at a premium, as for instance, in stationary storage applications, ultracapacitors could be used in combination with V2G and/or advanced stationary batteries like sodium-sulfur batteries, provided there are significant reductions in cost.

New developments in capacitor technology indicate the potential for these devices to move from a niche role in the energy system to a bigger role in energy storage. Nanotechnology may enable a large increase in the area of electrical charge storage in capacitors without increasing their bulk. Such devices are still being researched in laboratories and it is by no means assured that the indicated promise can be realized technically or, if it is, that the economics will be favorable. But that promise is important in the context of a renewable energy system.

Specifically, nanocapacitors (also called supercapacitors) have the potential to increase the energy density of capacitors 30 to 60 watt hours per kilogram.[98] While such capacitors would still be too heavy for most vehicular applications, they could serve as the basis for energy storage in small-scale renewable systems or as complements to a V2G system if they were cheap enough. That is a lot of ifs, and the potential may not be realized. This report does not rely on this technology in its scenarios. However, we have identified this as a research and development priority because the characteristics of nanocapacitors could enable a more efficient functioning of electric power grids and small-scale renewable energy systems.

Batteries can also be used for stationary storage. Specifically, the sodium-sulfur battery, which is bulky and unsuitable for transportation applications, can be used to store off-peak power generated by wind turbines.

3. Compressed Air Storage

Compressed air storage involves using off-peak electricity to compress air and store it in a large underground cavern, which could be a pre-existing cavern or one mined specifically for the purpose. At times of peak demand, the compressed air is withdrawn from the cavern, heated using natural gas, and used to operate a combined cycle plant. The advantage of this technology within this framework is that it can reduce the amount of expensive natural gas used per kilowatt hour and, in its place, use whatever fuel is available more cheaply at off-peak times. Design storage pressure can range from 1,100 to 1,500 pounds per square inch.[99]

The usual context for the use of compressed air storage in electrical power applications has been when cheap coal-fired capacity is used in the off-peak hours to compress air, but the approach can equally well be used for large-scale wind energy applications. There is less merit in this technology for central station solar technology, because solar energy already generates energy during peak or intermediate times. However, it may be useful for some hours of storage to provide electricity during the immediate post-sunset hours when electricity demand is still relatively high. Figure 3-16 shows a schematic of a compressed air energy storage system described above.

Figure 3-16: Compressed Air Energy Storage Schematic

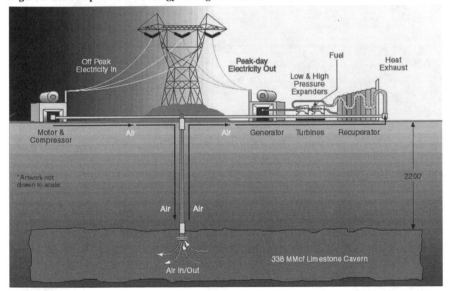

Source: Sandia National Laboratories

There are compressed air plants of medium size – one in Huntdorf, Germany (290 MW) and one in McIntosh, Alabama (110 MW). Both plants use salt caverns that were solution mined specifically for the purpose of providing compressed air storage for these facilities. The McIntosh plant has been in operation since 1991. It uses off-peak electricity to compress air and inject it into a compressed air storage cavern, and single stage natural gas turbines for on-peak power. Its cavern is 10 million cubic feet. Its nominal energy balance per kWh of peak output is as follows:[100]

- Off-peak use of 0.82 kWh of electricity from the grid to compress air – if this is coal-fired capacity, the fuel input would be 8,200 Btu.
- On-peak recovery of compressed air which is heated with 4,600 Btu of natural gas
- The combined result is 1 kilowatt hour of electricity during times of peak load takes 12,800 Btu of energy but 8,200 of that is cheap coal.

The overall energy balance is about the same as generating peak power with a single stage gas turbine. The result in the Alabama case is lower fuel cost but larger CO_2 emissions. At $7 per million Btu for natural gas and $1.25 per million Btu for coal, the cost of fuel is reduced by about 4.7 cents per kilowatt hour overall with the compressed air system. But the CO_2 emissions increase from about 680 grams per kilowatt hour for the single stage turbine to about 1,030 grams per kilowatt hour, an increase of about 350 grams emissions per kilowatt hour.

However, the same system can be deployed quite differently in the context of a goal of reducing CO_2 emissions. Specifically, compressed air storage can be used to store off-peak wind energy and displace single stage turbine use of natural gas. Since wind energy has essentially zero-CO_2 emissions (to a first approximation), the use of compressed air to displace single stage turbine use of natural gas with the same parameters as above (0.82 kilowatt hour of off-peak electricity and 4,600 Btu of on-peak natural gas) results in a net reduction of about 440 grams of CO_2 per kilowatt hour generated at peak, compared to using a single stage gas turbine without compressed air storage. A wind energy power plant combined with compressed air storage is being planned in Iowa.[101] In the long-term, that is, beyond 2030 or 2040, the natural gas can be replaced by methane made from biomass.

A great deal of optimization of large-scale wind, solar, and storage systems, including, possibly, compressed air systems would be necessary to arrive at a sound estimate of an economical combination of generation capacity (assuming only wind and solar were available) and compressed air storage. When one considers that baseload capacity in the form of geothermal energy and biomass fueled power plants will be part of the generating system in a zero-CO_2 economy, the scale, or even the necessity of compressed air systems that would be needed, is not clear. Since it is desirable for the electricity supply system to evolve as rapidly as possible in the direction of a reliable system based on renewable energy sources, further development of compressed air storage provides an important element of flexibility in actually achieving the goal.

L. Long-term Sequestration of CO_2

Coal used for electricity generation accounts for about one-third of U.S. energy sector emissions of CO_2.[102] The gravity of the global warming crisis has caused a considerable study of the technologies for capturing and sequestering CO_2 in underground or undersea geologic formations. A brief overview description of the approach is provided by Wilson, Johnson, and Keith:

> Geologic sequestration is accomplished by injecting CO_2 at depths greater than ~1 km into porous sedimentary formations using drilling and injection technologies derived from the oil and gas industry. The technology required to inject large quantities of CO_2 into geological formations is well-established. Industrial experience with CO_2-enhanced oil recovery (EOR), disposal of CO_2-rich acid gas streams, natural gas storage, and underground disposal of other wastes allows confidence in predictions about the cost of CO_2 injection and suggests that the risks will be low. Once injected, evidence from natural CO_2 reservoirs and from numerical models suggests that CO_2 can – in principle – be confined in geological reservoirs for time scales well in excess of 1000 yr and that the risks of geological storage can be small.[103]

The caveat "in principle" is important. As is generally recognized, a considerable amount of field research and development has to be done before the caveat can be removed and sequestration pursued with the necessary confidence that almost all of the confined CO_2 will remain confined for the long-term and that the potential for accidental large releases is acceptably small. A broad debate on the levels of demonstration that would be needed for widespread deployment has not yet happened.

In general, the types of geologic media that could hold large amounts of CO_2 are understood from prior experience, much of which derives from knowledge accumulated in the course of more than a century of oil and gas development and production. But it is necessary to have extensive measurements of leakage rates and rates of reactions of gaseous CO_2 with the surrounding geologic media to form solids in order to develop reliable models of long-term performance and estimate uncertainties. Figure 3-17 shows various methods of CO_2 sequestration (see color insert).

Saline reservoirs where CO_2 can form carbonates are considered to be among the most promising sequestration media. Such reservoirs also happen to be present in coal rich areas in the West, for instance, in Utah. A recent study by the Utah Geological Survey mapped the potential reservoirs in relation to existing sources of power plant CO_2 emissions. According to this study, the geologic formations "indicate [that] natural, long-term storage of carbon has occurred as precipitated carbonate minerals (mineral trapping) as well as by hydrodynamic trapping of gas and dissolved CO_2 in the pore water."[104] The potential for sequestration is indicated by the fact that about 100 million tons of CO_2 are generated by power plants close enough for the CO_2 to be piped into available geologic formations.

Modeling found that storage occurred in the gaseous, liquid, and solid phases. However, the solid precipitate is slow to form, so that containment of gaseous storage for several hundred years must be assured:

> The modeling suggests that there is ample storage in geologic structures beneath the Colorado Plateau, but a critical factor is whether the reactions that precipitate CO_2 have time to occur.
>
> These reactions typically require time scales of hundreds of years, so subsurface trapping for at least 500 years is essential. If major, high permeability faults are present, then loss of CO_2 to the surface could make the injection site unsuitable for CO_2 sequestration.[105]

The Utah Geological Survey model indicates that even after 1,000 years, the CO_2 would be well contained.

Much work remains to be done both in terms of commercialization of CO_2 capture and sequestration. The demonstration that the degree of containment required will endure for long periods of time will take considerable effort. At present not enough data are available for a confident conclusion. Yet, the scale of the use of coal in the United States and abroad is such that the development of the technologies and their demonstration is critically important.

In this study, the development of CO_2 sequestration is regarded mainly as a hedge – an element of flexibility that should be developed because:

- Coal is in widespread use and its use is likely to continue for some time
- Sequestration of CO_2 from biomass burning can provide for the negative CO_2 emissions that may become necessary if the actual impact of greenhouse gas emissions is greater than now projected
- Our approach to zero-CO_2 without nuclear power requires many different new technologies to work together and difficulties that are hard to foresee may arise, for instance, in the large-scale use of biomass or in the development of hot rock geothermal technology.
- Sequestration may also become very important if it is found necessary to re-move CO_2 from the atmosphere beyond zero-CO_2 emissions. In view of these considerations, the vigorous development of IGCC technology, CO_2 capture and sequestration is part of our recommendations, but actual continued reliance on coal and large-scale use of sequestration is not.

CHAPTER 4: TECHNOLOGIES—DEMAND-SIDE SECTORS

Here we take up the technologies and approaches in the energy consuming sectors – residential and commercial (considered together, since they are dominated by similar end uses), transportation, and industrial. Our analysis on the demand-side is first on the basis of delivered energy – that is energy that is actually used at the consuming site or in the consuming sector. The energy losses in electricity generation are separately considered.

A. Residential and Commercial Sectors

Residential use of energy is dominated by space heating, water heating, and space cooling (air conditioning). Figure 4-1 shows the energy use in the residential sector in 2004 – and these three end uses accounted for 56 percent of the total. But 46 percent of the total use of 21.07 quadrillion Btu was actually lost, discharged as waste heat at power plants, leaving just over half, 11.46 quadrillion Btu delivered to end users (Figure 4-2). On the basis of delivered energy, space heating, water heating, and space cooling combined dominate residential energy use, accounting for 71 percent of it.

Actually, a great deal of the delivered energy used for space heating is also wasted due to poor design of buildings and inefficient space heating systems. Therefore, most of the delivered energy used for space heating is wasted at the point of use. The same is true of water heating, since very high quality sources of energy, like natural gas and electricity, are used to produce hot water at very low temperatures. Most of the potential of the energy to do work is wasted when it is used for low temperature applications, for which other approaches such as solar water heating, are much more efficient.

Figure 4-1: Residential Sector Energy By End Use: Total Energy, Including Electricity Sector Losses, 2004.

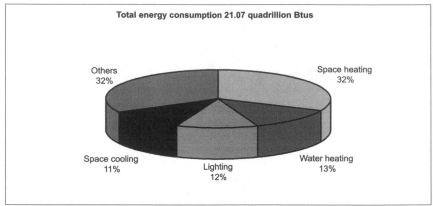

Total energy consumption 21.07 quadrillion Btus

Others
32%

Space heating
32%

Space cooling
11%

Lighting
12%

Water heating
13%

Source: EERE 2006 Table 1.2.3 (page 1-6)

Figure 4-2: Residential Sector Energy By End Use: Delivered Energy, 2004

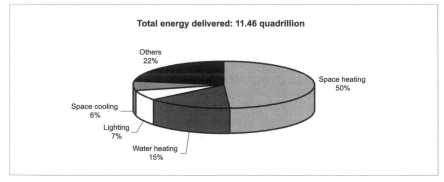

Total energy delivered: 11.46 quadrillion

Others
22%

Space heating
50%

Space cooling
6%

Lighting
7%

Water heating
15%

Source: EERE 2006 Table 1.2.3 (page 1-6)

The pattern is somewhat different in the commercial sector in that lighting is the largest single end use and water heating is not as important when losses in electricity generation are included (see Figure 4-3). This is, of course, to be understood in the context of offices, shops, etc., having a large lighting demand. Lights also heat up the air, increasing air-conditioning demand in the summer. In the winter, lighting reduces heating demand for the same reason. As a result of these factors, electricity use is high in the commercial sector and more than half (52 percent) of the energy use of 17.4 quadrillion Btu is discharged as waste heat at power plants. When only delivered energy is considered, space heating is the largest end user (Figure 4-4), but, as in the residential sector, a lot of that delivered energy is wasted in inefficient building design and heating systems.

Figure 4-3: Commercial Sector Energy By End Use: Total Energy, Including Electricity
Generation Losses, 2004

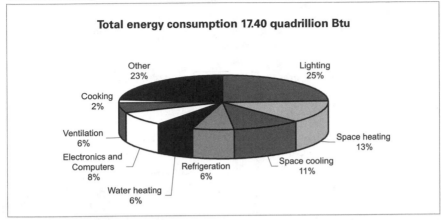

Source: EERE 2006 Table 1.3.3 (page 1-10)

Efficiency in lighting is critical to the performance of the commercial sector.
Including delivered energy plus waste heat in electricity generation, lighting is
25 percent of the total commercial sector energy use. When only the delivered
energy is counted, lighting is only about 16 percent of the total.

Figure 4-4: Commercial Sector Energy By End Use: Delivered Energy 2004

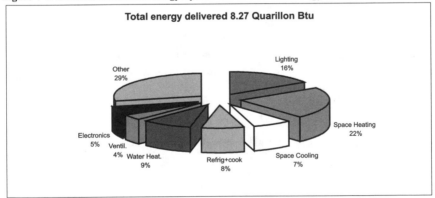

Source: EERE 2006 Table 1.3.3 (page 1-10)

The performance of the average building in the residential and commercial
sector can be classified as dismal compared to available technology and design
concepts, even leaving aside use of renewable energy sources. The main design
components and concepts have been known for some time:

- Excellent insulation
- Optimal thermal mass, designed for the climate – that is, a building that can

store sufficient heat on sunny winter days to be able to keep the home warm at night and on the next day if it is cloudy, but not so large that it would become too hot on consecutive sunny days

- Windows of sufficient area that let in heat and light in the winter – for starters, preferentially south facing (in the United States) – and can be shaded in the summer if necessary
- Very efficient lighting, appliances, and space heating and cooling systems.

If a solar water heating system is added to such features, most of the fuel requirements of residential buildings can be eliminated. The rest can be supplied in a variety of ways, depending on the overall cost of various energy sources and the policies in place at any time. Increasing lighting efficiency and use of sunlight directly and via special luminaires are especially important in the commercial sector. The actual achievement of excellent performance, within the parameters of a given set of energy prices and policies, will not always be reached, but it is worthwhile to examine what has been accomplished by sound design across the United States.

Below we describe two kinds of newly built residences, in two different climates. We compare the level of energy used in each of these buildings to the U.S. averages. One is a single family home in New Hampshire (Hanover House). The other is a multi-family apartment building with 43 units in Washington, D.C. (Takoma Village).[1]

The Hanover House in New Hampshire has a solar thermal water heater that provides both space heating and water heating. It has an electric hot water heater element that supplements the solar heat. There is a large storage tank. The use of solar heat keeps the electricity requirements for heating to a minimum. (Passive solar design by contrast uses the structure of the house to absorb heat, special windows, etc.). Its energy design features are as follows:

- **"Wall Insulation**
  Achieve a whole-wall R-value greater than 25
- **Solar Cooling Loads**
  Orient the building properly
  Locate garages and porches on the east and west sides of the building
- **Heating Loads**
  Site the building for southern exposure
- **High-performance Windows and Doors**
  Use superwindows with a whole-unit U-factor less than 0.25 (greater than R-4.0)
  Avoid divided-lite windows to reduce edge losses
- **Heating Systems**
  Use active solar heating

- **Air Infiltration**
    Use continuous air barriers
    Seal all penetrations through the building envelope

- **Computers and Office Equipment**
    Use Energy Star computer equipment"[2]

The only purchased energy input is electricity. Over a three year period, electricity consumption ranged from 4,250 to 5,560 kilowatt hours per year. The overall use of delivered energy was only about 8,300 Btu per square foot compared to about 58,000 Btu per square foot for the U.S. average in 2004.[3] The total energy, including electricity losses, was about 25,000 Btu per square foot for Hanover House compared to 109,000 Btu per square foot for the U.S. average. Overall there is about a factor of seven difference in the end use energy and more than a factor of four difference in the total energy.

Note that a 3 kilowatt solar PV system would be sufficient to convert this house to a zero net energy system. In that case, total energy would be reduced by a factor of 13 compared to the present residential average. Zero net energy homes with very low energy use have been built. An example in Arcata, California uses a geothermal heat pump, efficient building design and appliances, solar cooking (for 1/3 of the total cooking), and a 3 kW peak solar PV system.[4] Measured data over a four-year period show a small net electricity output (generation greater than consumption by 0.05%). Total electricity usage, including heating and appliances averaged only about 3,400 kWh per year.

A similar pattern emerges for multifamily housing. Note that Takoma Village Cohousing was a nearly completely commercial project, other than a $5,000 tax credit for first time home buyers among the residents. Washington, D.C. is hot and humid in the summer and moderately cold in the winter. Heating and cooling is provided by an earth-source heat pump (also called a geothermal heat pump). This gathers energy from the ground in a fluid that circulates in a buried pipe, which greatly increases the efficiency of the heat pump. A simple payback time of 9.5 years was estimated for the heat pump system.

The energy design features are:

- **"Wall Insulation**
    Minimize wall area through proper building massing
    Achieve a whole-wall R-value of 15 or greater
    Use spray-applied insulation in cavities with many obstacles or irregularities

- **Ground-coupled Systems**
    Use ground-source heat pumps as a source for heating and cooling

- **Solar Cooling Loads**
    Use light-colored exterior walls and roofs
    Minimize number of east and west windows
    Shade south windows with overhangs

- **Daylighting for Energy Efficiency**
  Use light pipes and/or active tracking skylights for daylighting

- **Non-Solar Cooling Loads**
  Reduce internal heat gains by improving lighting and appliance efficiency

- **Cooling Systems**
  Size cooling equipment appropriately
  Keep cooling equipment, especially air handlers and coils, in conditioned space

- **Foundation Insulation**
  Use slab perimeter insulation with an insulating value of R-11 or greater

- **High-performance Windows and Doors**
  Use windows with a whole-unit U-factor less than 0.49 (greater than R-2.1)

- **Heating Systems**
  Keep heating equipment in conditioned space

- **Luminaires**
  Use high-efficiency luminaires

- **Air Infiltration**
  Keep all mechanical, electrical and plumbing systems within the air and vapor barriers
  Perform blower door testing

- **HVAC Distribution Systems**
  Seal ducts
  Keep duct work out of unconditioned space

- **HVAC Controls and Zoning**
  Use seven-day programmable thermostats"[5]

The total end use energy was 26,300 Btu per square foot, with 21,100 of that being purchased electricity and the rest natural gas, compared to 58,000 Btu per square foot for the national average in 2004. Total energy use including electricity losses was 69,000 Btu per square foot, compared to the national average of 109,000 Btu per square foot.

A reduction of 60 to 80 percent in delivered energy (which is the point of reference here since the electricity supply system can change substantially) is easily possible in new construction. The technologies are well established.

Figure 4-5 compares the delivered energy use per square foot for the average U.S. house with the two examples discussed above.

Figure 4-5: Comparison of Two Efficient Homes with the U.S. Average Residential Energy Use (2004), Delivered Energy, Btu per Square Foot

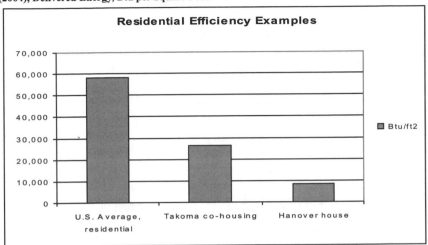

Source: IEER

The inefficiencies in the commercial sector are similar. For instance, the end use energy at the Durant Road Middle School in Raleigh, North Carolina, is about 25,000 Btu per square foot, and the total including thermal losses in electricity generation is 42,000 Btu per square foot. The comparable national averages are 103,000 Btu per square foot and 217,000 Btu per square foot respectively – differences of about a factor of four and five respectively. The design features responsible for the better energy efficiency were:

- **"Solar Cooling Loads**
  - Orient the building properly

- **Daylighting for Energy Efficiency**
  - Use south-facing windows for daylighting
  - Orient the floor plan on an east-west axis for best use of daylighting
  - Use north/south roof monitors and/or clerestories for daylighting

- **Interior Design for Light**
  - Use light colors for surfaces and finishes

- **Light Levels**
  - Use light levels appropriate for different tasks

- **Light Sources**
  - Use high-efficacy T8 fluorescent lamps

- **Lamp Ballasts**
  - Use automatic-dimming electronic fluorescent lamp ballasts in conjunction with daylighting

- **Luminaires**
  Use high-efficiency luminaires
- **Lighting Controls**
  Use on/off photoelectric daylight sensors"[6]

It is interesting to note that nearly all of the above features relate to lighting. Indirectly this would also reduce cooling loads and improve efficiency in the summer.

Consider an office building as another example: the Pennsylvania Department of Environmental Protection's Cambria Office Building in Ebensburg. It is an all-electric building with an earth-source heat pump. The end use energy is about 40,000 Btu per square foot, including 1,610 Btu per square foot of solar PV. In addition to its efficient heat pump and active solar energy, its design features include efficient lighting, insulation, high performance windows, etc.[7] For the commercial sector, it also appears possible, with existing design features, to reduce energy end use per square foot by three to four times compared to the present average. And neither example we have cited includes the use of combined heat and power. As with the residential sector, the technologies are well established. Figure 4-6 compares average energy use per unit area in the commercial sector with the examples discussed above, based on delivered energy.

Figure 4-6: Comparison of Two Efficient Commercial Buildings with U.S. Average Commercial Energy Use (2004), Delivered Energy, Btu per Square Foot

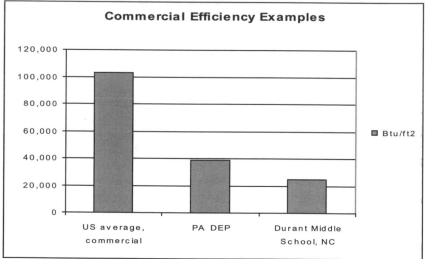

Source: IEER

The inefficiencies in the residential and commercial sectors provide key examples of the large-scale failure of the market and the resultant excess greenhouse gas emissions. A principal problem is that the developers generally do not pay the energy bills. This is called a "split incentive" barrier. The developer has an interest in the lowest capital cost possible compatible with building codes and sales strategy, while the occupants paying the bills have an interest, at least in theory, in the lowest overall annual operating cost (capital and energy bills combined). We will address this problem for new and existing buildings in Chapter 7.

It is worthwhile to mention some potential savings in appliances besides the well known potential in refrigerators and lighting. For instance, standby power consumption in a variety of devices like TVs and DVD players has grown to 600 kilowatt hours per household per year. These could be reduced to 200 kilowatt hours using 1 W or less standby systems.[8]

Backfitting, or retrofitting, existing homes is generally more complex than incorporating energy efficiencies in new buildings at the time of construction. Nonetheless it has been shown that many backfits can save energy and money when carried out properly. Consider, for instance, the case of a housing project of single-family houses for low-income households where backfits, such as better insulation and windows, were installed. There are measured data for this case, so that both energy performance and cost effectiveness were verified. The eight houses in this case study were in Florida.[9]

Backfits had short payback times. The shortest was one year – associated with cleaning refrigerator coils. Other measures – low flow showers, compact fluorescent lighting, and return duct sealing had payback times between 3.3 and 3.7 years. One house was backfitted with a solar water heater. This yielded the largest energy savings – 1,960 kilowatt hours per year. The payback time was estimated at 10.2 years. The electricity price used was a fixed rate of 8 cents per kilowatt hour.

A look at the change in the load profile, which is variation in the electricity demand over time, due to the solar hot water heater indicates that the economics would be dramatically different. Figure 4-7 shows the change in the load profile of the house backfitted with a solar water heater as measured between 1996 and 1998. There was a drop of about 500 watts in the peak load of the water heater. The solar water heater actually resulted in a reduction in load at most times of the day except for the period between 10 p.m. and 6 a.m. These are not the times of peak load for the utility, which are normally in the middle of the day or the early evening hours. Hence, there is a net benefit to the overall system that should be reflected in the costing of the program.

Another important result of the case study was that the payback time for the solar water heater installation in a new home was about the same as backfitting an existing home. However, the payback time was generally much lower for other devices if they were installed at the time of house construction. The biggest difference was the case of taping the duct system, which is much more laborious to backfit. Still, the payback time for a backfit was a respectable 3.6 years. When done properly in the first place, the payback time was only 0.7 years. These measured data, while sparse, are quite consistent with policies of building low-income housing to stringent efficiency standards and of backfitting existing housing so as to improve efficiency.

Figure 4-7: Load Profile of a Electric Water Heating System Without and With a Solar Water Heating Supplement

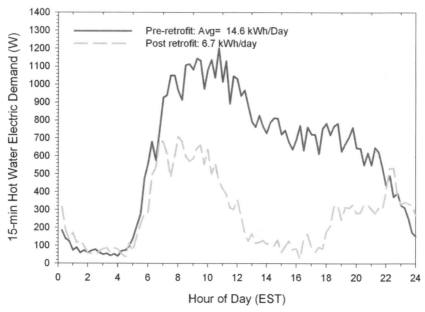

Courtesy of Florida Solar Energy Center. Source: Parker, Sherwin, and Floyd 1998

Lighting

Incandescent bulbs, which are still by far the most common type, typically convert two to three percent of the electrical energy input into visible light. This means that their efficiency on the basis of fuel input for electricity production is about 1 percent. This is because about two-thirds of the fuel input to coal and nuclear plants is discharged as waste heat at the power plant and the other one-third is converted to electricity and transmitted to the user. Compact fluorescent bulbs, which have been commercially available for some time, are about three to four times as efficient as incandescent bulbs and last much longer. One disadvantage is that, like other fluorescent bulbs, they contain mercury and the disposal problem has yet to be systematically addressed.

Emerging technologies beyond the compact fluorescent lamp have the potential to further reduce lighting energy use. Two examples are:

- Hybrid solar lighting: This technology uses optical fibers, which transmit sunlight from the outdoors to the insides of a building. They are in effect solar light pipes and conduct light much as a copper wire conducts electricity along its length with little leakage out the sides. A four-foot diameter solar concentrator on a rooftop that focuses light on a bundle of optical fibers is sufficient to provide light to about 1,000 square feet of indoor space at the height of a sunny day. The light pipes are part of lighting fixtures that also have electric lamps. As available sunlight increases or decreases, electronic sensors automatically adjust the light output of the electric lamps so as to keep overall light intensity constant.[10] The system was developed at Oak Ridge National Laboratory. It is being field tested in offices and large retail stores.[11]
- New LED lighting has an efficiency of 80 lumens per watt,[12] which is double that of compact fluorescent lamps.

One can anticipate that with such technologies, combined with motion detectors and photoelectric switches, electricity demand for lighting per unit area in many parts of the commercial sector might be reduced by about 80 percent (possibly more in some cases) in the next two decades. Electricity for residential lighting could be similarly impacted, notably since incandescent bulbs are still by far the most common in this sector.

B. Transportation

Figure 4-8 shows the end use pattern in transportation for 2004. Personal (light duty) vehicles and trucks are nearly four-fifths of the total and aircraft represent another 10 percent. The remaining ten percent miscellaneous set of items, while small, is critically important, since it includes everything from pipelines that transport oil and gas to barges that transport food grains to intra- and inter-city buses. Almost all the energy use in the transportation sector is supplied by petroleum. A tiny amount consists of electricity.

The problem of poor efficiency of personal passenger vehicles is well known – it arises from a combination of preferences for large vehicles on the part of consumers and aggressive marketing of such vehicles by manufacturers. While gasoline and diesel prices have fluctuated a great deal, the peaks that cause consistently high consumer demand for more efficient vehicles have not been sustained in the past.

We have already discussed electric cars and plug-in hybrids in the review of batteries, notably lithium-ion batteries (Chapter 3). The main problem at this stage appears to be large-scale manufacture and process improvements within the framework of the innovations that have already been tested and used in new vehicles, such as the Tesla Motors racing car, 0 to 60 in four seconds, the Phoenix Motorcars' pickup truck and planned SUV, as well as plug-in hybrids.[13] We assume that, with the right incentives, electric cars will become the norm in a reasonable time – twenty to thirty years. In the interim, we assume that plug-in hybrids will take a significant share of the institutional and then commercial markets, due to rising efficiency requirements, cost of fuels, and government and corporate procurement of advanced vehicles.

Figure 4-8: Transportation Sector Energy By End Use, 2004

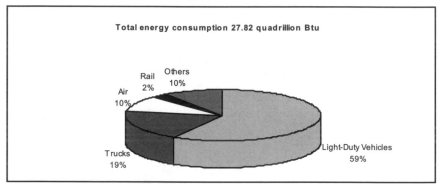

Source: EIA AEO 2006 Table A7 (page 145)

In this section on transportation technologies, we focus on fuels for jet aircraft (the predominant type of aircraft) and on the efficiency of public transportation.

1. Fuel for Jet Aircraft

For a zero-CO_2 economy, there are two basic approaches for replacing specially formulated kerosene (JP-8), which is the present fuel for jet aircraft. One can use biofuel feedstock to produce liquid biofuels, like biodiesel or ethanol or biofuel equivalents of liquid petroleum gases. Aircraft can also use fuels that are gases at room temperature provided they are liquefied. This requires cooling them to cryogenic temperatures. The fuels that have been studied are liquefied natural gas (LNG) and liquefied hydrogen. LNG can be replaced by methane made from biofuels.

Biodiesel and possibly some other liquid biofuels can, with some processing, be used in existing aircraft, though there remains considerable work to be done before a fuel has satisfactory performance and can be made at an acceptable cost. To use hydrogen fuel, aircraft would have to be redesigned to accommodate storage, because, for the same amount of energy, four gallons of hydrogen are necessary to replace one gallon of kerosene.[14] The issues relating to liquefied methane are on the one hand similar to biofuels, in that methane must be made from biofuel feedstock in a zero-CO_2 economy, and to hydrogen on the other, because a cryogenic fuel must be carried aboard aircraft. For simplicity, we discuss only liquid biofuels and hydrogen here, with the understanding that events may show cryogenic methane to be a preferable fuel. For instance, it has a higher volume energy density than hydrogen.

Biodiesel has some disadvantages as a fuel. The main one is that it freezes at a higher temperature than kerosene. Attempts to address this issue result in other problems, such as increased costs and lower fuel density. If a recent solicitation of bids by the Defense Advanced Research Projects Agency (DARPA) is any indication, it will take considerable time, effort, and money to produce biofuels suitable for jet aircraft at an acceptable price. The solicitation is quoted at length, since it provides many insights into the nature of the obstacles to be overcome:

> The Defense Department has been directed to explore a wide range of energy alternatives and fuel efficiency efforts in a bid to reduce the military's reliance on oil to power its aircraft, ground vehicles and non-nuclear ships. DARPA is interested in proposals for research and development efforts to develop a process that efficiently produces a surrogate for petroleum based military jet fuel (JP-8) from oil-rich crops produced by either agriculture or aquaculture (including but not limited to plants, algae, fungi, and bacteria) and which ultimately can be an affordable alternative to petroleum-derived JP-8. Current commercial processes for producing biodiesel yield a fuel that is unsuitable for military applications, which require higher energy density and a wide operating temperature range.... Subsequent secondary processing of biodiesel is currently inefficient and results in bio-fuel JP-8 being prohibitively expensive.

> The goal of the BioFuels program is to enable an affordable alternative to petroleum-derived JP-8. The primary technical objective of the BioFuels program is to achieve a 60% (or greater) conversion efficiency, by energy content, of crop oil to JP-8 surrogate and elucidate a path to 90% conversion. Proposers are encouraged to consider process paths that minimize the use of external energy sources, which are adaptable to a range or blend of feedstock crop oils, and which produce process by-products that have ancillary manufacturing or industrial value. Current biodiesel alternative fuels are produced by transesterification of triglycerides extracted from agricultural crop oils. This process, while highly efficient, yields a blend of methyl esters (biodiesel) that is 25% lower in energy density than JP-8 and exhibits unacceptable cold-flow features at the lower extreme of the required JP-8 operating regime (-50F). The focus of this program is to develop alternative or additional process technologies to efficiently produce an acceptable JP-8 surrogate fuel. Potential approaches may include thermal, catalytic, or enzymatic technologies or combinations of these. It is anticipated that the key technology developments needed to obtain the program goal will result from a cross-disciplinary approach spanning the fields of process chemistry and engineering, materials engineering, biotechnology, and propulsion system engineering. The key challenges are to develop and optimize process technologies to obtain a maximum conversion of crop oil to fuel....

> While the efficiency of the oil to JP-8 conversion process is the primary objective of this solicitation, the cost and availability of the necessary feedstock materials should also be considered. The development of conversion process technologies compatible with oils from a broad range of crops, potentially including new crop stocks selected specifically for their oil harvest, is preferred. Proposers will be required to provide a production cost model supporting their assertions of affordability.

It has been demonstrated that oil-producing crops (seeds and algae for example) can be genetically modified or selected to have certain desired agronomic characteristics, such as a higher yield of specific triglycerides. Proposers to the BioFuels program are encouraged to consider the use of selected crop oils (or mixtures) including specific cultivars, strains, etc., to maximize the conversion energy efficiency (crop oil to fuel)....

The program will be an exploratory evaluation of processing crop oils into a JP-8 surrogate biofuel, resulting in a laboratory scale production to be tested at a suitable DOD test facility. The successful proposer is expected to deliver a minimum of 100 liters of JP-8 surrogate biofuel for initial government laboratory qualification....[15]

Since a fuel that is not far from possessing the desired properties can be produced today, we have used jet fuel derived from biomass in the reference scenario. Hydrogen is also a possibility.

The commercialization of hydrogen fuel for aircraft will take considerable time and faces many uncertainties. Despite that, there are sound reasons to pursue research and development and further demonstration of the use of hydrogen as the standard aircraft fuel of the future. First, its technical feasibility has already been established in a commercial passenger jet. In 1988, the Soviet Union successfully demonstrated in flight a Tu-155 commercial aircraft that had been converted to use liquid hydrogen. It was also tested with liquefied natural gas in 1989.[16]

There are also strong arguments that, despite its poor reputation, hydrogen is a safer jet fuel than kerosene, though, of course, any accident containing a large amount of any flammable fuel is, by its nature, very dangerous.[17] Since hydrogen is a gas at quite low temperatures, it evaporates very rapidly upon release and, being much lighter than air, it disperses very fast. While liquid hydrogen needs a larger volume than jet fuel for the same amount of energy, it has a higher density per unit mass. The lower weight of fuel that would have to be carried could provide a significant boost in energy efficiency.

The European Aeronautic Defence and Space Company (EADS N.V.) has studied the feasibility, environmental impact, safety, and economics of liquid hydrogen powered aircraft.[18] A study by Airbus Deutschland in 2003 evaluated the prospects for hydrogen fuel in considerable detail. We use it here as a basis for the analysis of the prospects for hydrogen, especially as it is supported by other investigations. According to the study, which was based in part on a study of the performance characteristics of four conventional jet aircraft engines tested with hydrogen fuel:

This CRYOPLANE System Analysis has shown that hydrogen could be a suitable alternative fuel for the future aviation. Nevertheless, due to the missing materials, parts, components and engines further R&D work has to be performed until hydrogen can be used as an aircraft fuel. According to estimations made during this project the earliest implementation of this technology could be expected in 15 to 20 years, provided that research work will continue on an adequate level.

From the operating cost point of view hydrogen remains unattractive under today's condition, with kerosene is much cheaper as hydrogen and production/infrastructure is completely missing.

Assessments based on conservative calculations and today's understanding have confirmed that the use of hydrogen would reduce aircraft emissions to a minimum. It needs to be validated that the water emission of hydrogen-fuelled aircraft has low impact to the atmosphere as predicted.[19]

Airbus also estimated that "no technology leap is required" for hydrogen fueled aircraft.[20] In fact, according to Airbus Deutschland:

> This system analysis on components has demonstrated sufficiently that technology and design principles for H_2 fuel tank and H_2 fuel systems are available today....No showstopper for the further development of the CRYOPLANE has been found. However technical work has to be done in order to adapt and optimise the existing materials, components and modules to the needs of an aircraft design.[21]

The overall conclusions of the Airbus Deutschland study regarding a "realistic" time frame for commercialization of hydrogen fuel is surprisingly short – 15 years:

> Taking into consideration uncertainties both on the aircraft as well as on the infrastructure side a time schedule for having the first cryoplanes in regular airline operation can be estimated at approximately 10 (very ambitious) to 15 years (realistic).[22]

The main change in the aircraft would be in the configuration of the fuselage to accommodate the larger volume of fuel. The large volume of hydrogen fuel makes fuel tanks in the wings, which are used in kerosene-fueled aircraft, impractical.

Hydrogen-fueled aircraft would have lower environmental impacts overall than those fueled with petroleum-derived jet fuel. Large reductions in nitrogen oxide (NO_x) levels are possible; emissions of carbon monoxide and unburned hydrocarbons would be eliminated.[23] These advantages also hold for hydrogen relative to biofuels. There is one potential major problem relating to hydrogen, which is that it would produce more water vapor than jet fuel (and, in the future, biofuels).

Water vapor in the stratosphere is a greenhouse gas of some concern. Therefore the greenhouse gas emissions impact of a switch to hydrogen fuel depends strongly on the altitudes at which the aircraft would fly. Figure 4-9 shows a comparative evaluation of the overall greenhouse gas emissions of hydrogen and kerosene. At a 12-kilometer altitude (about 40,000 feet), hydrogen has about half the greenhouse gas impact of kerosene, but this is reduced to a very small fraction at 9 kilometers,[24] (about 30,000 feet). However, there is a fuel penalty, since the efficiency of jet aircraft increases with altitude.

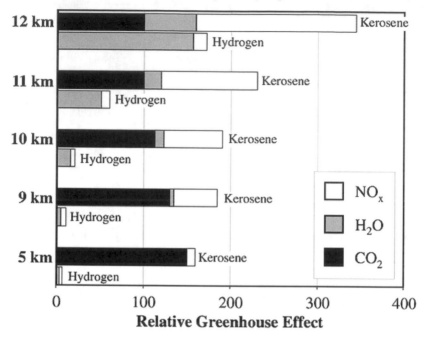

Relative Greenhouse Effect

Source: IPCC 1999, Figure 7-37 (Section 7.8.1) Used with permission.

As regards efficiency of aircraft, Airbus projects that jet fuel consumption as low as 1.5 liters or even 1 liter per 100 seat kilometers can be achieved.[25] The latter figure corresponds to over 230 seat miles per gallon. In this study we have assumed an average fuel efficiency of 150 seat miles per gallon by 2050.

2. Public Transportation

Excellent public transportation in cities is often one of the central features of making living in them convenient, and attractive. Paris and London and San Francisco are examples. Especially in cities with high traffic congestion on the roads, like Washington, D.C. or Los Angeles, with its attendant economic, environmental, and health impacts, there is a strong argument that people using public transport are subsidizing those using private cars, especially at times of peak travel, in more ways than one.

A good public transportation system is not only an important ingredient of livable cities, but it can save energy indirectly since fewer people choose to use their cars routinely in such cities. In many instances, they may own fewer cars or even forgo them. New York City is a prime example. It has the most diverse and efficient public transport in the country. It also has the lowest rate of vehicle ownership. As of the 2000 U.S. Census, less than 50 percent of households owned a car (all five boroughs). In Manhattan, fewer than one in four households had a car.[26] While this is to some extent a function of income (owner occupied households have greater vehicle ownership than renter occupied households), the existence of a diverse public transportation system is one critical element in overall low car ownership. Not coincidentally, New York City also has one of the lowest per-capita energy use rates in the United States, less than one-third of the U.S. average.[27]

The evidence on the energy efficiency of public transport is, as a general matter, more mixed. It is not a given that public transport is generally more efficient than personal cars. The efficiency of public transport is highly dependent on ridership. That in turn is dependent on density of cities, and the density and availability of public transport. Figure 4-10 shows the contrarian evolution of the efficiency of public transport buses compared to personal cars since 1970. The energy use per mile of cars has declined, while that of buses has increased.

Figure 4-10: Evolution of the Energy Use per Mile Versus Transit Buses Since 1970

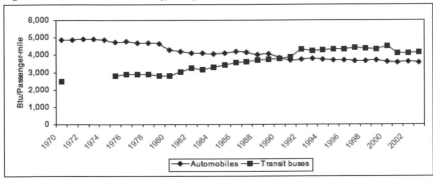

Source: TEDB 2006, Table 2-11

The reasons are not far to seek. First, personal passenger vehicles have had to comply with efficiency standards (known as CAFE or "Corporate Average Fuel Economy" standards). Despite the slippage in recent years, the improvement since the early 1970s, when car efficiency was typically in the 12 to 15 miles per gallon range, has been very large. Buses have not had to comply with such standards, and their fuel efficiency per vehicle mile has zig-zagged over the years rather than improved, while the efficiency per passenger mile has declined. Figure 4-11 shows the fuel consumption of transit buses per vehicle mile and per passenger mile (the inverse of efficiency).

Figure 4-11: Transit Bus Fuel Use per Mile: 1975 to 2003

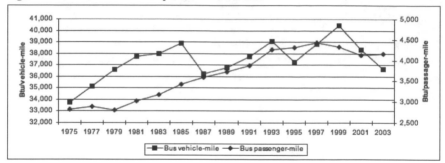

Source: TEDB 2006 Table 2-11[

The data in Figure 4-11 allow the computation of average ridership in a bus for a typical mile of its route. Figure 4-12 shows that ridership has declined since 1975 by about 25 percent.

Figure 4-12: Evolution of Transit Bus Ridership, 1975 to 2003

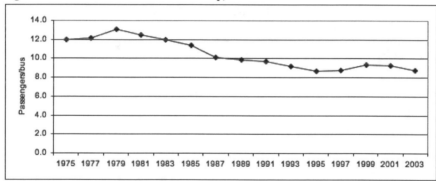

Source: TEDB 2006 Table 2-11

Declines in ridership, of course, make transit buses more expensive per mile to operate, creating a vicious circle of increasing cost, declining ridership and decreasing efficiency. A detailed investigation of the history of public transportation infrastructure is beyond the scope of this book. We only note here that the data indicate that the energy efficiency of public transport depends on whether and how well the system serves the public, whether it is affordable, and so on. A city that is well-served with public transportation will tend to have a more dense population, with lower car use and lower per person energy use. Figure 4-13 shows the estimated fuel consumption per passenger mile of three kinds of public transportation systems – light rail, buses, and heavy rail – in various cities.

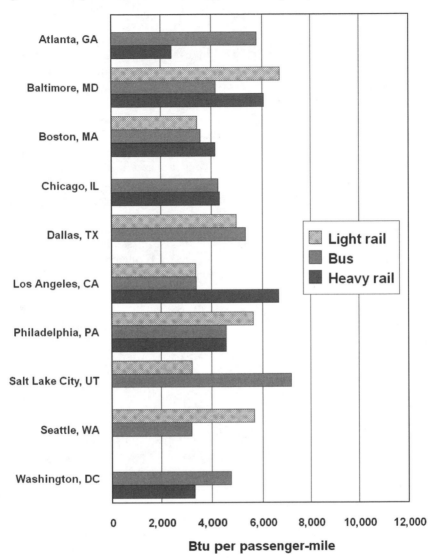

Figure 4-13: Comparative Efficiencies of Urban Public Transport Systems

Courtesy of Oak Ridge National Laboratory. Source: TEDB 2006 Figure 2.2 (page 2-15)

Efficiency does not appear to be a characteristic of the technical mode, but rather of other characteristics that are particular to the public's use of the system (including population density, service in the areas needed, etc.). The relatively high efficiency of the transit bus system in Los Angeles is perhaps one of the most interesting features of this chart. A demand for better public transport by the public of Los Angeles, notably its lower income public, and for economic and environmental justice were joined together in a long struggle that has resulted in Los Angeles becoming a surprising success story, still developing, in public transport.[28]

As a final note, we might consider the health benefits of living in a city in which walking, bicycling and greenways, public transport, mixed zoning and other considerations, are larger features than they typically are today in most U.S. cities and suburbs with segregation of housing, recreation, shopping, etc. An epidemiological study recently completed in New York City indicated that people living in neighborhoods where walking was easy and purposeful – such as stepping out to buy groceries or to go to a restaurant – had a lower body mass index than people in areas of New York City without easy access to public transport, mixed zoning, etc.[29] Public transport should be considered as a public utility in large cities, much like water and electricity supply and sewage systems.

Of course, living in densely populated communities is not everyone's cup of tea, and perhaps may be preferable at certain times of life than at others. The observation is offered here as an example of the kinds of considerations that should go into public policy decisions about public transport and its real benefits to the public. They in turn should help determine how public transport should be developed and costed. We have not quantitatively factored in public transportation changes into the scenarios in this study because of the complex nature of the problem. However, we do assume that the vehicular efficiency of transit buses will improve and that policies will be put into place towards that end.

C. The Industrial Sector

The industrial sector is the most complex of all the demand sectors due to the huge number of different industries and the diverse characteristics of energy use in them. For instance, mining, heavy manufacturing, metals production, chemicals, light industry, textiles, paper, and glass are all in one large energy sector. More detailed breakdowns are available, but an end use analysis from the point of technology and efficiency would take a multivolume treatise.

Fortunately, such an analysis is not necessary in the context of this study for two reasons. First, it is possible to aggregate the data by the major processes and end uses typical of broad classes of industry. Second, the policy approach chosen here, which is basically to make large users of fossil fuels pay for emitting CO_2 while reducing the total amount of emissions allowed each year, would automatically encourage industry to seek both ways to increase energy efficiency and to increase use of renewable energy. Hence, this sector does not require a detailed analysis. If the emission allowances are reasonable and decline in a predictable manner, the innovation and investment will shift towards reducing CO_2 emissions.

We briefly consider the kinds of areas in which industry will likely reduce CO_2 emissions. We include the use of feedstocks in industry, even though they are non-fuel uses of fossil fuels, for two reasons. First, many of the feedstocks eventually result in greenhouse gas emissions. Second, replacement of fossil fuels in all sectors, including industry, is an important part of ensuring that zero-CO_2 emissions are realized.

Among the uses of energy (including electricity) in industry are:

- Process heating, whereby the materials being worked on are heated, as for instance in the recycling of scrap iron and aluminum, the rolling of steel, and heating of chemicals to achieve the correct temperature for reactions.
- Production of steam for process purposes, which requires use of fuel in boilers.
- Electricity for driving machines, typically electric motors, but also diesel pumps and the like.
- Petroleum, liquid petroleum gases, and natural gas for feedstock uses.
- Reduction of ores to metal, as for instance reduction of bauxite to aluminum metal.
- Distillation.
- Heating, air conditioning, and lighting of buildings.
- Fuel for onsite generation of steam and electricity (combined heat and power).
- Lighting.

As noted in Chapter 1, there has already been a remarkable shift in industrial energy use patterns since 1973 due to a variety of factors, including fluctuating prices of energy, which have risen to quite high levels in some periods, innovations in processes, and the changing composition of industry. A cap on CO_2 emissions, if it is stringent enough, will convert the current trend of flat energy use in industry with rising production into a trend of declining energy use with increasing production. There are still many opportunities in industry for improving efficiency within the framework of available technology, such as efficient lighting and motors. But innovation will also play a role.

Industries and companies that have taken early action for a variety of reasons, including environmental protection, improving profitability, reducing uncertainties, and anticipating restrictions on greenhouse gas emissions, already indicate the large potential. We took a look at DuPont as a brief case study both because it has taken (and is taking) early action and because DuPont's Director of Sustainability, Dawn Rittenhouse, arranged for me to interview her and her colleague, John Carberry, for this report. A summary of that interview is in Appendix B.

In 1999, DuPont set a goal of reducing greenhouse gas emissions by 65 percent and actually achieved 72 percent by 2003. Most of this was in the form of reductions of halocarbon process emissions in manufacturing. DuPont has a target of further reduction of 15 percent based on 2004 emissions, with halocarbon and energy-related emissions being part of the achievement of the goal. In other words, DuPont is already accomplishing a major reduction in greenhouse gas emissions and a significant reduction in CO_2 emissions even without legislated restrictions.

In the interview, John Carberry discussed a few of the kinds of steps that would be taken in the context of a global goal of 60 to 80 percent reductions of greenhouse gas emissions:

> In the chemical industry CHP [Combined Heat and Power] is a big one.
>
> Another is replacing distillation – one alternative is modernization of processes so you don't have so many operations that involve distillation. Or it could be replaced by crystallization or membrane separation technologies, for example. Other areas are steam system management, insulation, powerhouse modernization, steam trap management. Optimization for first pass first quality yield is a big one – that is, make it correctly the first time. If you don't make it correctly, you have to recycle the product and make it again and you have wasted all the energy that was used the first time.
>
> Optimizing the manufacturing efficiency of your facility is another one. If you are in a standby hot mode, you use 60 or 70% of the energy anyway. So you want to run 100% of capacity 100% of the time. Then there is optimized process control and finding alternatives to grinding of solid materials – grinding is highly energy intensive.[30]

Further discussion on industry-related energy policy is in Chapter 7. In the reference scenario we assume that there will be approximately a one percent decline per year in absolute terms in U.S. industrial energy use between 2010 and 2050. The use of fuels for industrial feedstocks is assumed to be constant.

CHAPTER 5: A REFERENCE ZERO-CO$_2$ SCENARIO

In this chapter, we set forth a reference zero-CO$_2$ scenario going to 2050, at which time there would be no fossil fuels consumed and no nuclear power generated in the United States. Variations upon this reference case are considered in Chapter 6.

Zero-CO$_2$ emissions without nuclear power is an admittedly ambitious goal that would do nothing less than revolutionize the energy supply in the same way that petroleum and electricity did in the last century. There would also be considerable changes on the demand-side in that economic growth would be accompanied by slowly declining energy demand. However, the precedent of zero energy growth with significant economic growth already exists in the United States; it occurred in the 1973-1985 period (Chapter 1). It is also noteworthy that energy use declined slightly between 2004 and 2006, while GDP continued to grow at 3 percent per year.

The reference scenario also serves to illuminate constraints on renewable energy supplies, such as land for biofuels and the need for additional reserve capacity in the electricity sector in the case of wind and solar energy. The possible different time-scales for transitions are discussed in Chapter 6. The recommendations of the study are developed once the reference scenario and potential alternatives are discussed.

The reference scenario also serves to set forth the assumptions underlying the projected demand that serve to demonstrate the reasonableness of a delivered energy use of about 45 to 50 quadrillion Btu by 2050. (Electricity and biofuels production losses are separately considered.) One goal of the eventual set of recommendations is that there must be sufficient flexibility on the supply-side to meet a contingency of a somewhat higher or lower demand than forms the basis of the supply estimates here. The possible variation in the total energy figure is likely greater than that of delivered energy, since energy losses depend a good deal on the specific mix of types of electrical generation assumed and the extent of the role of liquid and gaseous biofuels and how they are produced.

A. Residential and Commercial Energy Use

The economic assumptions underlying the reference scenario and its derivatives are in the category of "business-as-usual." Some of the specific figures that are very important in analyzing the demand-side are set forth in Figure 5-1 for the residential and commercial sectors. The residential area is projected to grow from about 200 billion square feet in 2004 (the base year for these projections) to about 380 billion square feet in 2050. The number of households will increase from about 113.6 million in 2004 to 175 million in 2050.[1] This means an increase in the area per household of about 25 percent.

Commercial space is projected to grow as well. It is shown in Figure 5-1, but to a different scale (on the right of the graph). It is expected to increase by about two-thirds between 2004 and 2050.

The main loads – heating, cooling, and lighting – scale approximately as area. Others, such as hot water, would scale more according to population, whose rate of increase is slower. We do not scale the use of energy services by population, but do it rather by area, since this leaves room for new appliances and uses that would not be accommodated by a straight population-based projection.

Figure 5-1: Residential and Commercial Sectors, Projections of Floorspace, in Billion Square Feet

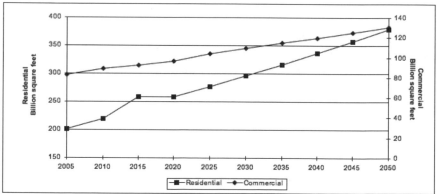

Sources: Commercial: DOE 2006 Table 2.2.1 (page 2-5) up to 2025. Residential: EIA AEO Assumptions 2006 page 23 and DOE 2006 Table 2.1.1 (page 2-1) EIA AEO gives an average square footage for 2001 and 2030. We have interpolated the values for the years in between and multiplied them with the number of households listed in DOE 2006. The values after 2025 for commercial area and after 2030 for residential area were extrapolated.

In estimating residential and commercial energy use and the efficiencies that can be achieved (using the approaches discussed in Chapter 4), we first calculate the energy actually used in the specific application. For instance, we derive a cooling load based on business-as-usual projections of efficiency and electricity use. These projections assume slow increases in efficiency not only for heating and cooling but for other appliances in the aggregate as well. For instance, the total heating load grows by only 10 percent and the cooling load by 40 percent, though the area almost doubles.

In the reference scenario the efficiency improvements are larger. There is a decline in delivered energy use from about 58,000 Btu per square foot per year in 2004 to about 21,000 Btu per square foot. In other words, delivered energy use per square foot would be about 37 percent of what it is today in the residential sector. We have shown by a few examples (and there are many more) that it is possible to design and build homes (single family and multi-family) that use between 8,300 and 26,000 Btu per square foot at reasonable cost in areas that are quite representative of conditions in large areas of the United States. Examples of even lower specific energy use can be found. Overall energy use on the basis of delivered energy would decline only about 30 percent, since the number of houses and the area per house are both expected to increase. Technology and efficiency assumptions are specified in the following endnote.[2]

Business-as-usual projections in the commercial sector actually assume an increase in delivered energy use per unit area, despite great potential for efficiency in new buildings. We have assumed that new space will be much more efficient beginning in 2015, but that existing space will achieve only modest energy efficiency increases by 2050. This recognizes that it is often more expensive to retrofit existing commercial buildings. Overall, energy use per square foot in 2050 would be about 58 percent of that in 2004, while total energy use in the commercial sector would stay about the same, due to increasing area. The technology and efficiency assumptions for the commercial sector are specified in the following endnote.[3]

Changes have also been assumed in the fuel supply of the residential and commercial sectors. We assume that most existing homes with natural gas as a heating fuel will convert to methane derived from biofuel, ordinary heat pumps, geothermal heat pumps, or resistance heating assisted by a solar thermal system (as in the Hanover House discussed in Chapter 4). Figure 5-2 (see color insert) shows the evolution of fuel and electricity use in the residential and commercial sectors combined, on a delivered energy basis. The transition from natural gas to methane can be expected to be smooth, since no changes in fuel transportation (pipelines) or infrastructure at the point of end use are involved. The efficiency slice is the avoided energy use due to increases in efficiency relative to the business-as-usual scenario.

B. Transportation and Industry

The personal passenger vehicle miles and aircraft vehicle miles in the business-as-usual projection are shown in Figure 5-3.

Figure 5-3: Business-as-usual Projections for Light Duty Vehicles (Vehicle-Miles Traveled) and Aircraft (Seat Miles Available)

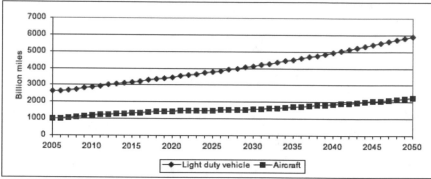

Source: EIA AEO 2006 Table A7 (pages 145-146) up to 2030, projected thereafter by IEER.
Note: Light duty vehicles are defined as weighing less than 8,500 pounds.

Cars that run on gasoline or diesel alone (including hybrid vehicles that cannot be plugged in) with efficiencies up to 60 miles per gallon that meet other safety and environmental standards, are available on the market today. Eighty-mile-per-gallon vehicles have also been manufactured. Plug-in hybrids can get 70 to 100 miles per gallon of liquid fuel; in addition, they use 0.1 to 0.15 kWh of electricity per mile. As is well recognized, much of the problem in the lack of use of highly fuel efficient vehicles has been the absence of stringent mandated efficiency standards, aggressive marketing of highly profitable SUVs, and customer preferences for the latter.

We assume gradual changes in new vehicle efficiency to 40 miles per gallon by 2020 and continued steady improvements after that to just under 75 miles per gallon by 2050, for liquid-fuelled vehicles. This yields an average fuel economy of about 65 miles per gallon in 2050.

The bigger change that is assumed here is a transition to steadily increasing use of electricity in light duty vehicles, until electricity dominates the energy input in this sector in about three decades. We envision that plug-in hybrids will first be introduced on a large-scale, followed by all-electric vehicles in about 20 years. These assumptions apply to the reference scenario. It is also possible that if direct production of hydrogen from solar energy and/or electrolytic hydrogen from wind energy become economical then a combination of hydrogen and electricity would be the mainstays for land transport. This possibility is discussed in Chapter 6.

Based on interviews and an examination of presently available data, which is scant, the present efficiencies of lithium-ion all-electric vehicles are in the 0.2 to 0.3 kilowatt hour per mile range (3.3 to 5 miles per kilowatt hour).[4] While there is an expectation that this will improve to 10 miles per kilowatt hour in the next several years, this appears rather optimistic. We have assumed an efficiency of 6 miles per kilowatt hour (delivered electricity at the plug) in 2015, slowly increasing to 10 or 11 miles per kilowatt hour in the 2040 to 2050 period for new vehicles made in that decade.

Partial use of electricity, in a mixture of plug-in-hybrid and all-electric modes is also assumed in commercial light trucks (50 percent by 2050), but the proportion of electricity for large trucks is small, 10 percent. This would account for a portion of the metropolitan area truck transport. We assume that developments in batteries will not be significant enough to allow long distance truck freight to be electrified.

There are fundamental reasons for seeking such a major transition in transportation technology and putting policies into place to ensure that it will occur:

- Electricity provides the greatest flexibility in energy supply.
- Use of solar and wind energy to charge plug-in hybrids and all-electric vehicles will greatly reduce waste of energy and increase transportation efficiency. With an efficiency of 5 miles per kWh, which is possible today, the use of solar or wind energy would yield an equivalent "well-to-wheels" efficiency of about 150 miles per gallon. This can be doubled in the coming decades.
- Making the transition to electric vehicles, for the most part, eases the pressure on other, more difficult, sectors, like aircraft and feedstocks in industry. The requirements of other sectors, combined with continued use of liquid fuels in industry, could put intolerable pressures on land for producing biofuels if passenger vehicles continue using liquid fuels.
- Electricity for transportation greatly reduces fuel cost, especially if the charging is mostly done off-peak. Hence, a greater investment in the vehicle itself is possible, for the same per mile transportation cost.
- The change would make the air in cities dramatically cleaner than it is today, since petroleum-fueled vehicles are the largest source of air pollution in many urban areas and, as such, are a principal contributor to respiratory diseases, like asthma, especially among children and the elderly.[5]
- New battery technology permits vehicle-to-grid support for renewable energy sources at nearly zero-marginal cost in terms of battery wear. This makes a V2G supported grid much more feasible and obviates the need for costly storage technologies. It also provides some insurance against difficulties in large-scale development of hot rock geothermal technology and other baseload sources to support a wind and solar PV system. Lithium-ion batteries can be recycled.

Figure 5-4 (see color insert) shows the evolution of the transportation fuel mix in the reference scenario. Initial uses of electricity are mainly for plug-in hybrids. The high efficiency of electric cars means that a relatively small amount of electricity can replace a much larger amount of gasoline. The energy use is shown on the basis of delivered energy; neither electricity production losses nor biofuel production losses are shown. They are discussed in Section C.

It is possible that technological developments in areas such as solar hydrogen production or hydrogen production from high-yield biomass, could turn out to be more economical than electricity. These possibilities are discussed in Chapter 6 as variants of the reference scenario. Rapid and large-scale introduction of plug-in hybrids into the marketplace could probably be achieved if they became a significant part of governmental and corporate fleets.

Tesla Motors is founded on the idea that initial market breakthroughs occur at the high-end of the market, since the wealthy are willing to pay more for an avant-garde, attractive all-electric car that is also environmentally friendly. At about $100,000 per car, the Tesla Roadster is already sold out for the 2007 model year and more than half of the 2008 model year has been reserved.[6] By design, the approach is similar to the introduction of new appliances and gadgets, such as digital TVs and cameras, DVD players, or, long ago, color TV, where the initial buyers were people willing to pay high prices, opening the way for cheaper mass manufactured products that displaced the prior standard ones.

Finally, as noted in Chapter 4, the reference technology for aircraft is continued use of the present type of jets with biofuels, with incremental efficiency improvements to 150 seat miles per gallon by 2050. Today's most advanced passenger commercial aircraft perform at about 100 seat miles per gallon.[7] The main technology and efficiency assumptions for the transportation sector in the year 2050 are discussed in the following endnote.[8]

Even with a very fundamental transition to electric vehicles for passenger vehicles and light duty trucks, transportation fuel requirements for aircraft and internal combustion engines remain very large – about 6 million barrels a day of oil equivalent in 2050. These requirements would by themselves be well within reasonable land requirements for production of liquid biofuels.[9] However, the industrial biofuel requirements must also be taken into account. They increase land requirements considerably.

We have assumed that energy use in industry for fuel uses will decline by 1 percent per year and still sustain business-as-usual growth in output. Feedstock uses of fuels would remain constant over time. Overall, this requires only a modest change from no-growth in energy use that has prevailed on average since 1973. The net result is that industrial energy use in 2050 would be about 70 percent of that in 2004 (delivered energy basis). This is a reasonable concomitant of an assumption of a CO_2 emission reduction regulation system in which emission allowances for large users will be fixed ("capped"), with the limit declining each year until it reaches zero by mid-century (see Chapter 7). An interview with DuPont officials on industrial energy use in a world with CO_2 emission restrictions is in Appendix B. DuPont is one of the corporations that is part of the U.S. Climate Action Partnership (USCAP),[10] which advocates, among other things, a target of 60 to 80 percent reductions in U.S. greenhouse gas emissions by the year 2050.

C. Electricity Production

About half the electricity production in the United States in 2005 was fueled by coal. About 19 percent came from nuclear energy and 19 percent from natural gas (including combined heat and power generation in industry). The balance came from hydroelectricity, petroleum, and renewable sources such as wood waste and wind. Solar-generated electricity was not yet a significant component of the supply.

Since over 90 percent of the generation came from thermal power plants, mainly coal and nuclear, the losses of energy were considerable. The overall generation efficiency of these two types of power generation, on average, is about one-third, which means that about two-thirds of the energy input winds up as waste heat. Since this waste heat component is a very large part of total energy use, it important to consider how it is actually accounted for in energy data. Without a careful consideration of this issue, energy data over time could be rendered non-comparable.

1. Methodological Note on Thermal and Other Losses in Electricity Production

Electricity by its nature is thermodynamically different than fuels that are burned to produce heat. In theory, electricity can be converted with 100 percent efficiency into mechanical energy (or work). The same is true of converting the mechanical energy in the flow of water into electricity. Heat energy conversion to mechanical energy (or electricity) is restricted to an upper limit less than 100 percent, determined by the temperature of the combustion relative to ambient temperature. The efficiency of thermal power plants is highly variable in practice. It ranges from a low of 15 or 20 percent for geothermal energy to about 33 percent for nuclear power plants, about 40 percent for new coal-fired power plants, and 55 percent for natural-gas-fired combined cycle plants. This has created a methodological problem. Electricity from all of these sources is equivalent, and after it enters the grid, its source cannot be determined. But hydropower needs no fuel. So how is the mechanical energy input to a hydropower plant to be added to the fuel input to a coal-fired or nuclear power plant? Assuming a unit of hydroelectricity is equivalent to a unit of coal used in a coal-fired power plant would be adding up incommensurate kinds of energy in terms of the useful work that can be extracted from them.

Traditionally, a fictitious heat loss, typical of thermal electricity generation, is added to hydroelectricity generation to make its contribution commensurate with fossil fuels. This creates an artificial inflation of energy use in an economy that does not correspond to actual energy use, since hydropower plants do not have such thermal losses. However, the practice does not result in a large distortion of energy data so long as non-thermal electricity generation sources are a small part of the total, as they are today in the United States. However, in a transition to an economy where wind and solar photovoltaic electricity would play a major role and where the efficiencies of thermal generation could range from 15 percent to 55 percent combined cycle plants, the traditional approach is quite unsuitable since it would greatly distort the actual energy inputs into the economy.

In this book, we have projected delivered energy, including electricity consumed at the point of delivery. That is, the basic analysis on the demand-side discussed above is done according to the evaluation of energy used at the point of use – homes, office buildings, cars, factories. On the supply-side, a variety of choices can be made for electricity generation, some of which would involve thermal losses, while others would not. For instance, a large role for biomass combustion would mean greater thermal losses than if some of that role were taken up by solar PV. The approach, therefore, is to produce scenarios of electricity supply that would meet the criteria of reliability, resource availability, and constraints (such as land), and then estimate the actual thermal losses that would result from the specific mix of sources.

These considerations are quite important in comparing different supply scenarios. The delivered energy remains the same in all cases.

In addition to thermal losses at the power plant, all centralized electricity generation entails losses of electricity between the point of generation and the point of use. These are called "transmission and distribution" losses. The term "transmission losses" applies to high-voltage electricity transmission from the generation plant to intermediate voltage points of use for large-scale industrial and commercial users or to substations where the electricity is converted to the low voltages that are typically used in homes, office buildings, schools, shops, etc. Distribution losses are from these intermediate points to residences and other small-scale uses. Large industries often take their electricity at higher voltages and do not have distribution losses. Overall transmission and distribution losses amount to about eight percent of electricity generation, with most of that being distribution losses. In the reference scenario, we have assumed that electricity losses go up slightly (from eight percent to ten percent) due to a greater use of the distribution system and lower use of the high-voltage transmission system. The losses could be reduced if generation at the point of use is increased.

2. Electricity in the Reference Scenario

The demand sector projections discussed above show electricity as part of the delivered energy to each sector. A transition to an electricity sector based on renewable energy sources requires a complex set of considerations. The first is reliability. The present electricity sector is highly centralized, apart from a modest amount of combined heat and power generation in the industrial sector (about 4 percent of the total). By and large, this provides a reliable supply, though its vulnerabilities have been apparent in various major blackouts in the past several decades, including the major Northeast blackout in 1965 and the most recent one in 2003.

These vulnerabilities stem from the potential for disturbances created by the removal of a major generating station or an important segment of the transmission grid at a time of heavy load. This can cause temporary disturbances in the grid, called transients, that cause more and more generating stations and/or sections of transmission lines to shut down for safety reasons (to protect against overloads). Blackouts can spread with great speed. It is a complex and difficult exercise to turn the entire grid back on after a widespread blackout. Many types of institutions, from hospitals to banks, have emergency power supplies that allow them to keep operating at minimal levels during blackouts. Nonetheless, prolonged blackouts lasting a few days cause immense economic damage and create health risks as well.

In addition to the risks of blackouts due to natural disasters (such as hurricanes and lightening strikes), excessively centralized systems are also vulnerable to terrorism, for the same reason. An attack on critical sections of the system could cause the same types of dislocation and damage as a prolonged blackout due to other causes.

On the other hand, a purely decentralized system also has its problems of reliability. A breakdown could cause a prolonged period without electricity, though the damage is restricted to a local area. For that very reason, a decentralized system presents a far less attractive target for terrorist attack than a centralized system. However, a purely decentralized system that is also reliable is generally expensive because extensive back up is required in case the main system is down for maintenance or due to accidents or natural disasters.

A mix of the two approaches with decentralized sources providing a large fraction of electricity connected into a grid that also has centralized sources can overcome most of the vulnerabilities of each approach. In fact, it can provide a more reliable system. A grid within which small-, intermediate-, and large-scale generating stations all play significant roles is called a "distributed grid." Distributed grids can also bring dispersed wind resources into the energy system in a much more cost effective way than a purely decentralized system, especially in the United States, where the best land-based wind energy resources are concentrated mainly across a swath through the middle of the country and offshore.

The total electricity requirements under the reference scenario remain about the same throughout the period under consideration (to 2050). Efficiency improvements reduce demand; this is offset by loads growing due to increasing economic output, greater numbers of homes and businesses, and new uses of electricity (such as plug-in hybrids and all-electric vehicles). But the fuel mix of electricity would have to change almost completely, except for the eight percent or so that comes from hydroelectricity, wind, wood wastes, and geothermal energy.

As we have discussed in Chapter 3, solar and wind energy are each plentiful enough to supply the entire electricity requirement of the United States. We have also discussed various ways in which the intermittency of these two resources can be addressed by optimizing their contribution to electricity generation based on overall cost for a given reliability.[11]

Besides combining wind, solar, standby natural gas/bio-methane, and hydropower to overcome the effects of intermittency, the reference scenario assumes the use of a V2G system after 2030 or 2035; in the alternative, stationary storage in advanced batteries, possibly in combination with ultracapacitors, can also perform the same function.

In order to provide baseload power, we assume a significant use of solid biofuels for electricity production, about 9 quadrillion Btu per year, generating over one-fifth of the total electricity requirement in the year 2050. The use of solid biomass is coupled to the production of microalgae from the CO_2 exhaust. This forms the feedstock for producing liquid fuels for transportation. In addition, methane derived from biomass would be used in combined cycle plants in place of natural gas in order to provide reserve capacity in the system. Hot rock geothermal power is also assumed to be deployed on a significant scale after 2030. This technology is important since it can provide baseload generation in areas that have relatively low solar energy availability and relatively low potential for large-scale biomass production at high efficiency, as for instance the Northeast.

Finally, the number of combined heat and power systems would grow in the industrial and also the commercial sector (with more modest use in the residential sector, for instance in multi-family housing). Natural gas is the main fuel for such systems today; it is assumed that this will be gradually replaced by methane made from biofuels.

Figure 5-5 (see color insert) shows the evolution of the electricity sector in the reference scenario. Solar energy consists mainly of solar PV, but also includes 150 gigawatts of solar thermal with heat storage for 12 hours. In this arrangement, solar thermal can serve as a kind of quasi-baseload generating system if built in very sunny areas such as the Southwest. The preferred technology for solid biofuels would be IGCC because of its efficiency and the relative efficiency with which CO_2 can be captured in this system. In the initial 2010-2020

Chapter 3

Figure 3-1: Colorado Green Wind Farm

Courtesy of DOE/NREL, Credit: Sandia National Laboratories

Figure 3-2a: Population Density

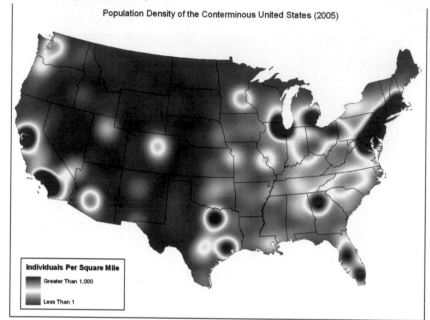

Provided by AWS Truewind, LLC

Figure 3-2b: Wind Resource Density

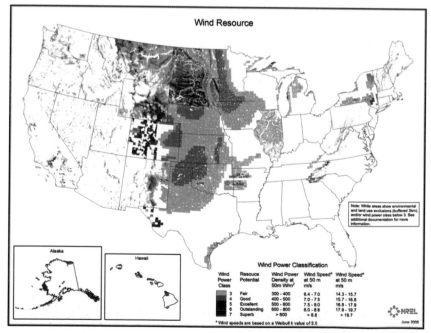

Provided by National Renewable Energy Laboratory.

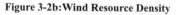

Figure 3-6: Solar Insolation, in kWh Incident per Day (Annual Average Values)

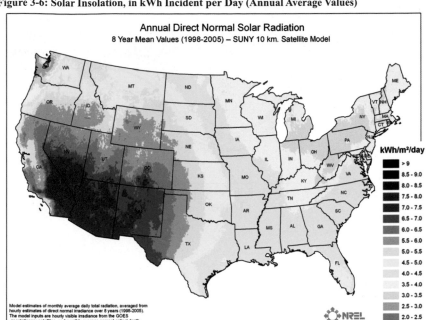

Annual Direct Normal Solar Radiation
8 Year Mean Values (1998-2005) – SUNY 10 km. Satellite Model

Provided by National Renewable Energy Laboratory.

Figure 3-7 Monthly Values of Available Insolation for the Equator, 30°, 60°, and 90°North

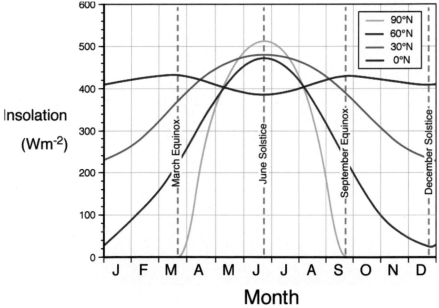

Created by Michael Pidwirny. See Pidwirny 2006 and www.physicalgeography.net

Figure 3-8: Net Power Bought: Average Hourly Profile-Zero Energy Home

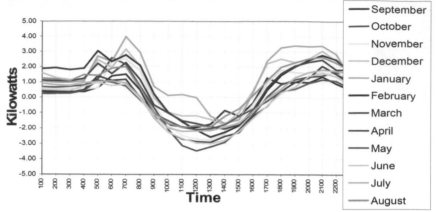

Courtesy of Environmental Resources Trust, Inc.

Figure 3-12: Water Hyacinths Can Yield up to 250 Metric Tons per Hectare in Warm Climates

Courtesy of Center for Aquatic and Invasive Plants, Institute of Food and Agriculture Sciences School, University of Florida

Figure 3-15: Direct Solar Production of Hydrogen Using Algae

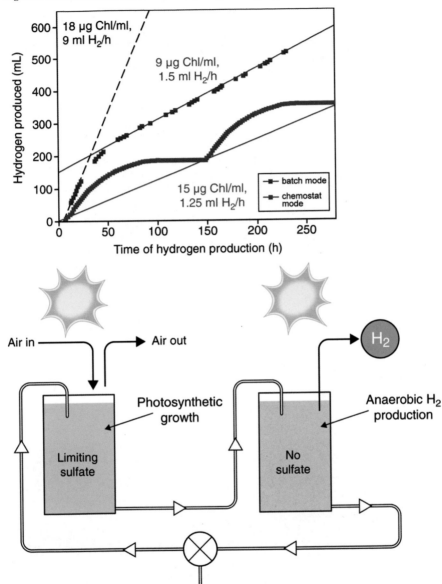

This diagram/graph was developed by the National Renewable Energy Laboratory for the U.S. Department of Energy. See Ghirardi and Seibert 2003.

Note: In the *"batch mode"* the production is stopped periodically to replenish the nutrients. In the *"chemostat mode"* nutrients are supplied continuously to maintain production. "Chl" stands for chlorophyll.

Figure 3-17: Schematic Showing Different Methods of CO$_2$ Sequestration

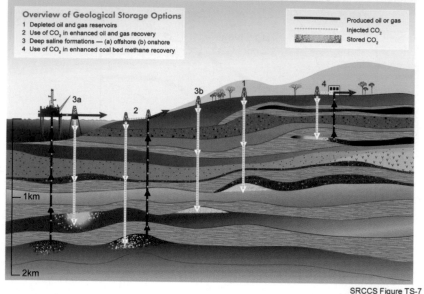

Methods for storing CO2 in deep underground geological formations

Source: IPCC 2005 Figure TS.7 (page 32). Used with permission.
Note: Airhart 2006 provides a good summary of sequestration.

Chapter 5

Figure 5-2: Residential and Commercial Energy, Delivered Energy Basis, IEER Reference Scenario

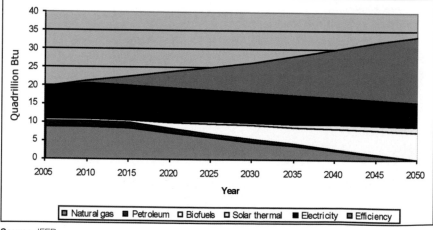

Source: IEER

Figure 5-4: Transportation Energy Use, IEER Reference Scenario

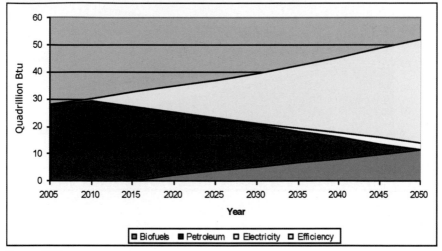

Figure 5-5: Electricity Supply, IEER Reference Scenario

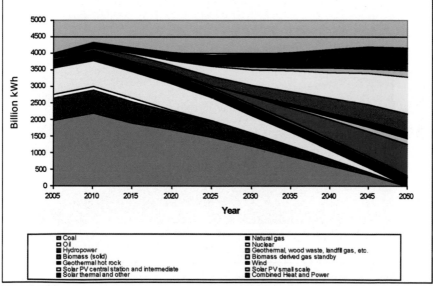

Figure 5-7: Delivered Energy, IEER Reference Scenario

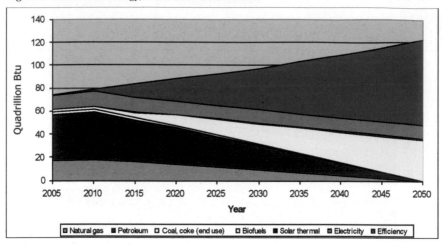

Source: IEER
Note: Fuels used for electricity generation are not shown here. See Figure 5-5.

Figure 5-8: Total Energy Inputs in the Transition to a Zero-CO$_2$, Non-nuclear Economy by 2050, IEER Referene Scenario

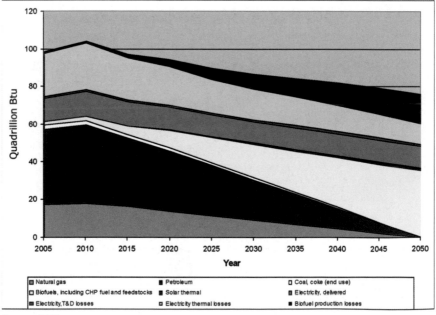

Source: IEER

period, a larger part of the renewable expansion is due to an increase in supply from wind energy. Of course, in this period, most of the present baseload capacity would continue to be available. We also assume that the use of CO_2 capture in microalgae would be implemented at existing fossil fuel power plants, so as to minimize emissions and create an industrial base for biofuel production that does not rely on food crops.

Much of the solid biomass would likely be prairie grasses or switchgrass. We will explore various alternatives for biomass production for electricity generation and of the use of solar energy for producing transportation fuels (other than electricity) in Chapter 6.

Figure 5-6 is a schematic diagram of the electricity system in the reference scenario. The numbers are similar to those in the reference scenario, but ranges are shown in some cases, for purposes of illustration. Other combinations are possible with this same set of technologies. The actual evolution of electricity supply will depend on relative costs, the state of transmission and distribution, infrastructure, and other factors.

Figure 5-6. One Possible Future U.S. Electric Grid Configuration Without Coal or Nuclear Power in the Year 2050

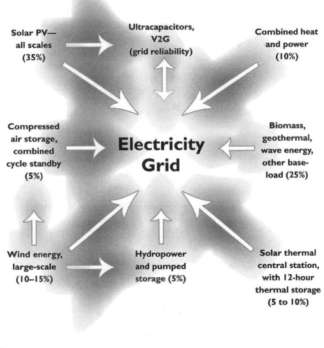

Source: IEER

In the scheme shown in Figure 5-6, about 45 to 50 percent of the electricity supply would be from intermittent renewables, not including solar thermal power
plants. This would require a considerable standby capacity, but not equal to the
peak demand. A coordination of wind, solar PV, and solar thermal in a way that
takes advantage of the diversity of times when they are available would reduce
standby requirements. A large portion of the standby would be supplied by combined cycle plants operating first on natural gas and then on methane derived
from biomass. There is ample spare capacity available and a good portion of that
would be maintained. Some standby capacity would be provided by hydropower.
Solar thermal power plants would be provided with 12-hour heat storage, so that
they could provide power through much of the time when bright sunshine is not
available. Further, about 25 percent of the capacity would consist of central station baseload or quasi-baseload capacity.

A combination of a V2G system and stationary storage, for instance, in advanced
batteries, would provide the rest of the backup. It is difficult to estimate what
this amount would be without developing detailed load profiles, which is far beyond the scope of this study. It would be less than and probably much less than a
quarter of the peak demand in the configuration shown in Figure 5-6.

We assume for the sake of estimation that the standby capacity required to be
supplied by a combination of V2G, advanced battery, and ultracapacitor storage
in the year 2050 would be on the order of 100 gigawatts, which is about equal to
the installed capacity of all U.S. nuclear power plants. This seems rather large,
but a very small fraction of the light duty vehicles would be able to meet it. At
10 kilowatts per vehicle,[12] the number of vehicles required would be 10 million. This is about three percent of the fleet of light duty vehicles in the United
States projected for the year 2050. Typically, vehicles are used much less than
10 percent of the time, so that on average over 90 percent of the vehicles would
in principle be available. However, a far smaller number of vehicles would be
available at peak vehicle use times. This will likely not have a significant effect
since only a few percent of vehicles would be required, at most. Hence, arrangements made with businesses that have large numbers of vehicles in their parking
lots at the time of peak load would be sufficient to provide adequate standby
capacity. Vehicles parked at airports could also play a role.

Storage of electricity on the supply end can be combined with storage equipment
at the demand end. For instance, an air-conditioning system that is equipped
with an ice-making machine can shift air conditioning load from on-peak times
in the middle of the day to off-peak hours. It is commercially available from Ice
Energy for both residential and commercial buildings.[13]

Such a system can complement renewable energy storage systems by shifting
the load to times when renewable energy is available. For instance, ice can be
made at night when wind energy is typically more available and used for air

conditioning during the daytime. Similarly, the peak of solar energy availability is in the middle of the day, while the peak of the air-conditioning load often occurs in the late afternoon.[14] Michael Winkler has proposed a "smart grid" system in which thermal storage (of both heat and coldness) is controlled by the utility to some extent so as to match available supply. In this concept, renewable energy sources, geothermal heat pumps, storage of heat and coldness, and electricity storage are combined so as to optimize the generation capacity and make the best use of available intermittent resources. A smart grid would allow greater use of intermediate- and small-scale solar energy with greater reliability per unit investment and potentially at lower cost.[15]

D. Overall Results

A series of graphs illustrate the results of this analysis. Note that generally we have assumed that major changes will begin between 2015 and 2030 depending on the state of the technology. Figure 5-7 (see color insert) shows the delivered energy in the reference scenario. The electricity shown in the chart is that actually consumed at the point of end use (rather than at the point of transformation to another energy source). Similarly, thermal losses and biofuel production losses are not shown. The increases in efficiency incorporated into the scenario result in a decline of delivered energy use from about 74 quadrillion Btu in 2004 to about 48 quadrillion Btu, a reduction of about 35 percent.

Figure 5-8 (see color insert) shows the total energy input into the system including electricity transmission and distribution losses, thermal losses in electricity production, and biomass losses in liquid and gaseous biofuels production. The total energy use declines from almost 100 quadrillion Btu in 2005 to about 76 quadrillion Btu. The losses in the present system are concentrated in the electricity generation sector. By contrast, in the reference scenario in 2050, the electricity system losses would be cut by more than half. However, the losses in production of liquid and gaseous biofuels for all end-use sectors will likely be large; as a result, the overall losses do not change significantly when comparing the energy system in 2004 to the reference scenario in 2050. The proportional role of losses in the renewable energy system in the reference scenario is actually greater than at present (almost 37 percent compared to 25 percent). This is undesirable. Alternative approaches are discussed in Chapter 6. These are used to develop a preferred renewable energy scenario (Chapter 8, Section A).

Land Use Considerations

Wind energy takes up relatively little land. Crops can be cultivated and cattle can graze right up to the towers of wind turbines, whose footprint is small. The area requirements for wind energy are determined by the swept area of the turbine blades, which does not significantly impact the footprint of the installation. For instance, the total footprint of 15 wind turbines, 2 megawatts each, in a Polish wind farm was only 0.5 hectares (1.25 acres). The project was built on an area totaling 225 hectares of farms. Almost all the land between the wind turbine tower foundations will be farmed.[16]

The largest area requirements are for the service roads associated with the construction and maintenance of wind farms. Other service facilities, such as an electrical substation, would also be required. The actual area required is site dependent, since the length of the roads would depend on topography, existing land uses, and other factors. An analysis by the New York State Energy Research and Development Authority concluded that five percent of the total land area of the project might be considered as rule of thumb for planning wind power projects. The total land-area requirements per unit of installed capacity themselves vary from project to project, and depend largely on the wind speed characteristics and topography of the site. Assuming a total project area of about 12 hectares per megawatt, the land-area requirements would be about 0.6 hectares per megawatt.[17] On this basis, the total land-area requirements for wind energy in the reference scenario would be about 490 square miles, which is equal to a square about 22 miles on the side.

Solar photovoltaic cells also do not take up much land. In fact, installations on rooftops and parking lots take up no additional land. Assuming that half of the large- and intermediate-scale installations are associated with commercial parking lots and rooftops, the land-area requirements for solar PV in the reference scenario are rather modest – about 860 square miles, which is equal to a square about 29 miles on the side, assuming the central station installations are in sunny areas. This includes a 30 percent allowance for roads, space between the PV arrays, and infrastructure.

We estimate solar thermal electric power production land requirements would be about 210 square miles. The trough or parabolic reflectors that track the sun in such power plants capture solar energy much more efficiently than solar PV, though much of that advantage is lost in the thermal electricity production cycle as waste heat.

Overall, the total land-area requirements in the reference scenario for wind and solar energy (other than parking lots and rooftops) would be about 1,560 square miles, which is a square almost 40 miles to the side.

Liquid and gaseous biofuels, derived from solid biomass grown for the purpose, play a very large role in the reference scenario. In fact, their role in the energy sector would be somewhat greater (proportionally speaking) than that played by oil and natural gas in the United States economy today. This is mainly because there is a very large component of industrial demand and a significant component of demand in each of the other sectors that cannot easily be met by electricity at reasonable cost, given present technology. The overall requirement for liquid and gaseous biofuels in the reference scenario is about 35 quadrillion Btu of delivered energy. This does not include solid biomass requirements for baseload electricity production or the losses associated with production of liquid and gaseous fuels from solid biomass. As can be seen from Figure 5-8, these losses are substantial. The total solid biomass production requirements for all uses in the reference scenario are about 60 quadrillion Btu. We have assumed an efficiency of 70 percent for liquid and gaseous fuels production from solid biomass by the year 2050.

A part of this requirement can be met by recovering landfill gas, which has a significant amount of methane (the principal constituent of natural gas). Gasification of household waste, use of waste cooking oils, and other sources can also provide some sources of fuel. However, a complete elimination of fossil fuels would create very large requirements for liquid and gaseous fuels, unless there is a transition to a hydrogen economy and/or a far greater use of solar thermal energy and/or electricity for a variety of purposes including space heating and industrial process heat. That is the case in the reference scenario. For purposes of illustration of land requirements in the reference scenario, we will ignore the relatively modest contributions that landfill gas and household garbage and trash could make to total biofuel requirements. In practice such sources can often be used to good effect.

The productivity of land and the efficiency with which the biomass is converted into liquid and gaseous fuels (mainly methane to replace natural gas) and feed-stocks determine the land area that will be needed. The use of prairie grasses and switchgrass for producing the entire projected amount would require 12 to 15 percent of the land area of the United States, which is an unrealistic requirement. Even if it were feasible, devoting such a large land area to commercial crops would require the creation of a vast new infrastructure of roads and industries in many areas that are now unspoiled or nearly so. For reference, the land area harvested in 2005 was 321 million acres,[18] which is about 14 percent of the U.S. land area.

The reference scenario, therefore, requires the inclusion of a substantial portion of high productivity biomass to reduce the land-area requirement to about 5 to 6 percent. The latter figure is the upper limit of what would be feasible (though not necessarily desirable). Six percent of the land area of the United States is about equal to the land area of Montana and North Dakota combined.

The principal ways to reduce land-area requirements while still relying on liquid and gaseous biofuels derived from biomass is to maximize the use of landfill gas and other waste biomass and to rely on biomass that has high efficiency of solar energy capture (~ 5 percent). The approaches are discussed in Chapter 3 and can be summarized in the context of the reference scenario as follows:

- Capture of CO_2, notably in microalgae, in the short and intermediate (5 to 30 years) term from fossil fuel combustion at power plants and in industry.
- Capture of CO_2, notably in microalgae, in the intermediate and long term (from about 2020 onwards) from biomass and liquid and gaseous biofuel combustion at power plants and in industry.
- Cultivate high productivity biomass, including microalgae and aquatic plants, such as water hyacinths and duckweed, for instance, in constructed wetlands associated with wastewater treatment systems and in areas with runoff that have high nutrient content.

The following approach has been used in the reference scenario regarding capture of CO_2 in the biomass/biofuels sector for the year 2050:

1. Twenty percent in industry
2. Fifty percent in production of liquid and gaseous biofuels from biomass
3. Eighty percent in central station electricity production.

The low percentage of CO_2 capture assumed for industry is due to siting issues, since land availability would likely be a problem for a large number of industries. This would be the smallest constraint for power plants, since these would be sited close to the location of biomass production, with due consideration given for land requirements of CO_2 capture in microalgae. The percentage of CO_2 captured from the liquid and gaseous biofuels production sector is assumed to be in between the industrial and power generation sector. In most of these cases, facilities for one-to-two-day storage of CO_2 would be required in order to capture the CO_2 generated at night on the following day or two. This would be required to accomplish the targeted capture fraction.

The productivity of microalgae and aquatic plants is assumed to increase from 150 metric tons per hectare (60 metric tons per acre) in the year 2020 to 250 metric tons per hectare in the year 2050. As noted in Chapter 3, the largest productivity that has been observed to date has been 250 metric tons per hectare under optimum climatic conditions.

With these assumptions and a productivity of switchgrass or prairie grasses of 30 metric tons per hectare by 2050, the land-area requirements for all biofuel requirements, including those for electricity generation come to about 184,000 square miles, which is just over 5 percent of the land area of the United States. It should be noted that these calculations of land area are very approximate and depend greatly on a variety of assumptions about the kinds of plants that would be grown, and the regions where the biomass would be grown.

Table 5-1 summarizes the main land-area requirements for the reference scenario:

Table 5-1: Land-Area Requirements for the IEER Reference Scenario (rounded)

Energy source	Land area, square miles	Side of a square	Comments
Wind	490	22	Mainly infrastructure, including roads
Centralized Solar PV	860	29	PV area + 30% infrastructure
Solar thermal (central station)	210	15	Collector area + 30% infrastructure
Biofuels (solid and liquid)	184,000	429	About five-sixths of the area is harvested area for biomass; rest is microalgae and aquatic plants
Total	**185,360**	**431**	**About 5.2 percent of U.S. land area**

Notes: 1. Wind capacity factor = 30% and land per megawatt = 0.6 hectares.
2. Solar PV efficiency 15%; average annual insolation 250 W/m^2.
3. Solar thermal efficiency 20%; average (tracking) insolation 300 W/m^2.

It is easy to see that the land-area requirements are dominated by biofuel production. This is because:

(i) the amount of biofuel requirements are very large, since biofuels supplant coal, oil, and natural gas combined, albeit in a more efficient economy,

(ii) the losses involved in the production of liquid and gaseous biofuels are significant even with overall 70 percent efficiency,

(iii) a significant amount of biomass production is assumed to occur at a rather low solar energy capture efficiency of 30 metric tons per hectare, which is an efficiency of solar energy capture of less than one percent at typical average levels of insolation.

Cultivation and harvesting of biomass must be done in ways that do not decrease the carbon stored in the soil (a minimal requirement) or, preferably, it should increase carbon stored in the soil. In this analysis it is assumed that biomass cultivation will not change soil CO_2 storage.

The reference scenario incorporates features that would allow land currently not deemed fit for cultivation and, potentially, as well as, areas such as the Salton Sea in California for most biomass cultivation. The land-area requirements are still very large. Cultivation of prairie grasses, switchgrass, etc., would require an expansion of harvested area in the United States by about 30 percent. If sufficient high productivity biomass is not available, the land-area requirements could increase beyond 6 percent. It is therefore important to consider ways to reduce the land-area requirements, including increasing biomass production efficiency and direct solar hydrogen production. We note here, in closing, that the reference scenario is designed mainly to illustrate one path to a zero-CO_2 emissions economy without nuclear power. It is not necessarily the most desirable way to get there. We explore the options in Chapter 6.

CHAPTER 6: OPTIONS FOR THE ROADMAP TO ZERO-CO_2

The reference scenario provides one plausible way to achieve a U.S. economy without CO_2 emissions or nuclear power by about 2050. However, on the basis of the technical framework in that scenario alone, there are a number of uncertainties that may prevent its achievement. It may also not be the most effective or environmentally sound way to a renewable energy economy. We have already noted the rather large land requirements (over 5 percent of the U.S. land area) for biofuels as well as the large energy losses associated with the production of liquid and gaseous biofuels in the reference scenario. Further, the continued use of carbon-based fuels also implies the continuation of some level of air pollution, including unburned hydrocarbons, carbon monoxide, and nitrogen oxides.

Further, several of the key technologies in the reference scenario are leading-edge technologies that are still in the demonstration stage, as for instance, is the case with capture of CO_2 from power plants using microalgae. Other technologies are in the marketplace, but are not yet commercial and require subsidies or cater to niche markets. This is the case with lithium-ion electric cars/SUVs, for instance. Lithium-ion batteries must come down in cost by a factor of about five before they can be used on a large-scale to transform the energy system. This is also a requisite for their use in an effective vehicle-to-grid system. The path to the zero-CO_2 emissions goal would be quite uncertain unless there is a systematic technological redundancy built into energy policy so that roadblocks in one or a few areas do not prevent overall progress towards eliminating CO_2 emissions.

A. Hydrogen Production from Solar and Wind Energy

It is possible today to produce hydrogen on a large-scale from renewable energy sources by electrolysis of water.[1] Hydrogen can be produced on a distributed basis, that is, near the point of use, or on a centralized basis. In the latter case, a hydrogen infrastructure, notably long-distance pipelines are needed. We will focus on distributed generation in this brief examination in order to illustrate the potential of hydrogen to displace biofuels.[2]

Figure 6-1 shows a flow diagram of a distributed hydrogen production system. It consists of an electrolyzer, water supply, a water purifier (since high purity water is needed), a compressor, a storage tank, and ancillary facilities. Vehicles can be refueled from the storage tanks. The overall efficiency of present-day systems was estimated to be about 60 percent as of 2005.

Figure 6-1: Schematic Diagram of Compressed Hydrogen Production by Electrolysis

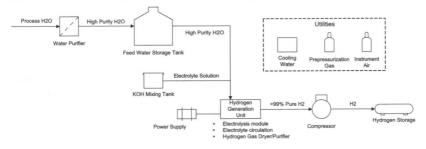

Source: Ivy 2004 Figure 1 (page 6)[3]

While a considerable amount of attention has been devoted to cars with fuel cells that use hydrogen as a fuel, this is not necessary for using hydrogen in motor vehicles. It can be used in present-day internal combustion engines. Pound for pound, hydrogen carries about 2.7 times as much energy as gasoline. However, since it is very light, its volume energy density is correspondingly low. Hence for cars to have a reasonable range, it must be compressed to 10,000 pounds per square inch or be used in the form of liquid hydrogen. The latter carries significant cost penalties.

A BMW luxury car prototype, with a 260-horsepower engine, that is fueled by liquid hydrogen, is being made in a limited edition, to be driven by selected users, on lease or loan in Europe, Asia, and the United States. A few liquid hydrogen refueling stations will be open to serve the drivers. The range of the car on hydrogen fuel will be limited to 125 miles. It is a dual-fuel car, with a supplementary gasoline fuel tank, which extends its range to 425 miles.[4]

The Department of Energy's program plan for hydrogen estimates the cost of distributed hydrogen production using electrolysis at about $4.80 per kilogram. The DOE cost estimate assumes an electricity cost of 3.9 cents per kWh, which is a low off-peak cost. This is a cost estimate not for wind-generated electricity, but rather among the lowest prevailing prices available on U.S. electricity grid.[5] Were the analysis done for wind-generated electricity, the cost of hydrogen would be higher – closer to about $6 per kilogram. This is double the average price of gasoline in the United States as of early July 2007 (energy content comparison). However, it is typical of the price of gasoline in much of Western Europe, since gasoline is highly taxed there.

The DOE estimates that in order to bring the cost of hydrogen to about $2.80 per kilogram, electrolyzer costs per kilogram of hydrogen would have to decline by about a factor of four from $1.20 in 2006 to 30 cents. Operating and maintenance costs, other than electricity, would have to decline from $1.40 to $0.70 per kilogram of hydrogen. A modest reduction in electricity costs from $2.20 to $1.80, mainly attributable to increases in electrolyzer efficiency, is also assumed to occur within a decade. With typical wind energy costs, these figures would imply a cost of about $4 per kilogram of distributed hydrogen production.

The above comparisons have treated hydrogen and gasoline on a par for the purposes of fuel cost evaluation. However, tests on prototype hydrogen cars using internal combustion engines indicate that their efficiency will be higher than the same cars using gasoline. A Ford 350-Series pickup truck using hydrogen was "up to 25 percent" more efficient than its gasoline counterpart according to a Ford hydrogen vehicle technical leader.[6] If a hydrogen car is significantly more efficient than a gasoline car, all other things being equal, then the break-even price of a kilogram of hydrogen can be that much higher than a gallon of gasoline. For instance, if hydrogen is 25 percent more efficient than gasoline, then hydrogen at $4 per kilogram is equivalent to gasoline at about $3.20 per gallon, if the pickup truck has a gasoline fuel efficiency of 15 miles per gallon. Further, hydrogen from renewable energy would have no CO_2 emissions and it would also have lower emissions of other pollutants than gasoline-fueled cars. The significant health benefits from reduced urban air pollution by switching to hydrogen fuel are not easy to quantify but very real.

As an aside, it is worth noting that electrolysis of water also generates pure oxygen ($2H_2O \rightarrow 2H_2 + O_2$), which could in some cases be marketed. If the electrolysis facilities are near a coal-fired power plant, the oxygen could be used instead of air for the combustion of coal. This would reduce nitrogen oxide emissions and enable capture of CO_2 for sequestration. We have not explored the possible implications of this, since it would require site-specific study, but considerations relating to the use of oxygen should be part of any optimization strategy for producing electrolytic hydrogen.

The Department of Energy's cost goals for electrolytic hydrogen discussed above are for the year 2017. If they are met, it may be possible to avoid much of the use of biofuels assumed in the reference scenario, since hydrogen could be used in its stead, possibly from 2025 onward. The reference scenario use of biofuels in the year 2050 for transportation excluding aircraft is about 9 quadrillion Btu. If half of this is replaced by distributed hydrogen, the land-area requirements could be reduced by 10 to 15 percent.[7] However, this would require quadrupling of wind energy requirements compared to the reference scenario.

The transmission infrastructure requirements would be very large and present a significant obstacle. Were wind-derived hydrogen to become economical, it would be possible to consider special pipelines for hydrogen. As an alternative, wind-derived hydrogen could be used to a more modest extent and coupled with direct solar hydrogen production. A mixture of wind-derived hydrogen and centralized direct solar hydrogen production could be considered. This would make the hydrogen infrastructure more economical, since it would be shared between wind and solar hydrogen production. This would improve the capacity utilization by reducing the impact of intermittency of either source alone.

Centralized hydrogen production would require a pipeline infrastructure, which could at least in part follow existing electricity transmission corridors. Such corridors already exist from the Midwest eastward and from the Rocky Mountain states westward. In addition or as a substitute, offshore wind farms could be used to create onshore distributed hydrogen infrastructure. Offshore wind farms may be the best approach in many cases to combining large-scale wind energy with distributed hydrogen production, since the wind farms could be built within a few dozen miles from the points of hydrogen production on land.

Hydrogen could also be used for residential and commercial applications in place of biomass-derived liquid fuels or methane. In an economy in which most biofuels are replaced with hydrogen produced at 10 percent efficiency from solar energy, the land requirements for a renewable economy could be reduced to ~2 to 3 percent of the US land area – or about half that of the reference scenario. Wind-derived hydrogen would take even less land. We note that 10 percent is currently the DOE target efficiency for photoelectrochemical hydrogen production for the year 2018. This is a method of producing hydrogen directly from solar energy (see Chapter 3).

B. Efficiency and Electricity

It is possible to reduce biofuel requirements in the residential and commercial sector by increasing efficiency relative to the reference scenario and, in that context, also increasing the use of electricity.

In the reference scenario, the average residential energy use per square foot is about 38 percent of the average in 2004. For the commercial sector the value is about 58 percent. There are a many energy efficient buildings being built today, some of which are not much different in cost than less efficient ones that have energy use significantly less than the projected average. The Hanover House, a single family home already discussed in Chapter 4, is an example. The delivered energy in 2004 on average was 58,000 Btu per square foot in the residential sector and that in the Hanover House was only 8,300 Btu per square foot. A combination of advanced design features and active solar thermal hot water and space heating minimize the purchased energy. One of the most interesting results of this design is that the house uses no liquid or gaseous fuels at all. The supplemental heating is provided by electric resistance heat. The combination of active solar thermal heating and design features means that even a rather inefficient use of electricity – resistance heating – is in a context where the inefficiency of the method is rendered more or less irrelevant due to the small demand. As noted in Chapter 4, the house could achieve net zero energy with about a 3 kilowatt peak solar PV installation.

While it is not possible to backfit existing homes with all the features of the Hanover House, it is possible to backfit many more existing homes with space solar heating and possibly solar thermal cooling as well.[8] One of the principal advantages would be to largely eliminate methane derived from biomass. A detailed evaluation of the potential for residential and commercial use of such technologies both in existing and new buildings would provide a guide as to the amount of methane replacement for natural gas that can be eliminated.

As another example, we have used an average coefficient of performance of six for air conditioners in the year 2050 and of four for heating for geothermal heat pumps in that year. The best current commercially available equipment using geothermal heat pumps has a coefficient of performance for cooling of about eight (Energy Efficiency Ratio or EER of 27) and heating of about four.[9] A gradual increase in standards to a cooling COP of eight or ten and a heating COP of five or six is likely possible, with the right incentives and regulations.

In the transportation sector, efficiency of liquid fuel use can be pushed considerably beyond that assumed in the reference scenario. For instance, the efficiency of light-use vehicles (personal cars and SUVs) is assumed to increase gradually to 50 miles per gallon by 2027. By contrast, the European Union has a target of 52 miles per gallon by 2012. Of course, the United States is far behind the EU currently, so that it will take time to catch up. But there is little reason, other than political resistance by the automobile industry in the United States, that the efficiency schedule in the reference scenario cannot be accelerated to 50 miles per gallon by 2020 and 100 miles per gallon by 2050. The increases in efficiency of trucks can be similarly accelerated.

Aircraft in the reference scenario also have slow improvement in efficiency, which on average would reach about 100 seat miles per gallon by about 2035. This efficiency has already been achieved by current generation of new aircraft. With an average life of aircraft in service at any time of ten years, a much greater improvement in efficiency is possible and perhaps likely, given current high fuel costs.

Finally, it is also possible that reduction in battery cost and weight would allow electrification of long distance truck transport. This is a matter whose evaluation can properly be done in a few years, when battery technology is more mature and prototypes have been built as they have been for cars and light trucks.

Overall, the liquid and gaseous biofuels requirement could be cut possibly by roughly a third, possibly more, with present and easily foreseeable efficiency standards and incentives as well as greater orientation towards electrical and solar thermal heating technology.

C. Stationary Storage of Electricity

It is possible that vehicle-to-grid approaches would not work as well in practice as the promise indicated on paper. In some circumstances, combinations of high peak loads and low availability of vehicles at the right locations may make reliable operation difficult. It is appropriate therefore to consider the cost of stationary storage. This can be done using advanced batteries (lithium-ion, sodium sulfur) possibly with ultracapacitors.[10] The latter can be considered for supporting the electricity grid but not for cars (so far as can be foreseen) because they store much less energy per unit weight than do lithium-ion batteries or even lead acid batteries. Since weight is at a premium in vehicles, batteries are to be preferred for electric cars. That is not a critical constraint for stationary applications.

Lithium-ion or other advanced batteries, possibly in combination with ultracapacitors, could be used to provide storage for solar PV systems as a complement to or in place of V2G if the overall capital cost of storage is reduced to $200 per kWh or less. The added capital cost of one day's storage, including ancillary equipment, would be about $1,200 per installed peak kW of solar PV capacity.[11] At $1,500 per peak installed kilowatt for solar PV, the overall cost of electricity provided at peak and intermediate times works out to about 16 cents per kWh. Distribution costs for electricity generated on an intermediate-scale in commercial parking lots or on commercial rooftops might be on the order of 2 cents per kWh. With a more efficient use of electricity, the overall cost of electricity services (lighting, refrigeration, air conditioning, etc.) would not be significantly different than at present (see Chapter 8). It appears worthwhile therefore to place a significant emphasis on developing stationary storage methods for electric power with a cost goal of $200 per kWh or less.

D. Feedstocks and Industrial Energy

A very large use of liquid and gaseous fuels (at present oil and natural gas and, in the reference scenario, liquid and gaseous biofuels) is for use as industrial feedstocks, as for instance for plastics, lubricating oils, synthetic textiles, and other products, such as vehicle tires, made from synthetic fibers. Feedstock uses of energy-containing materials are projected to remain constant at somewhat over 7 quadrillion Btu per year through to the middle of the century. This is about one-fifth the estimated use of liquid and gaseous biofuels in 2050 in the reference scenario – about the same as the entire use of these fuels in the residential and commercial sectors combined.

Recovery of materials for reuse where they may be burned or discarded today would be a much more powerful incentive in the context of policies designed to eliminate CO_2 emissions. Fossil fuel feedstocks would be treated on a par with fuels since most such materials eventually degrade and produce greenhouse gases, including CO_2. While some do so slowly – others, such as plastics and tires – are often incinerated. For instance, if a technology for devulcanizing rubber **can be commercialized** – that is the process that removes sulfur from rubber – then most raw material for new tires could come from discarded existing ones.[12]

In some cases, plastics can be recovered for replacing new feedstock. New materials can be designed that would ease such recovery. It is difficult to estimate the impact that such approaches would have cumulatively without a detailed study devoted to this subject alone. That is one of the reasons that they have not been included in the reference scenario. However, it would be highly desirable to reduce the use of feedstocks as much as possible so as to reduce the requirements for biofuels.

E. Natural Gas Combined Cycle and Coal as Contingencies for the Electric Grid

The electricity sector discussed in Chapter 5 relies a good deal on advanced technology such as lithium-ion batteries, the vehicle-to-grid system, and hot rock geothermal that are on the cutting edge of new developments in energy today. Technical assessments available today indicate that all of these technologies can be made economical within ten to fifteen years or less in the context of policies designed to achieve a zero-CO_2 economy (that is policies that increase the price of fossil fuel use and encourage the use of renewable energy and higher efficiency). But that is by no means assured. It is prudent therefore to make a contingency plan in case some of these approaches do not work. Direct solar hydrogen production as well as electrolytic production of hydrogen from wind are two such technologies. The latter is well in hand and requires a cost reduction of about a factor of two (compared to a factor of five for lithium-ion batteries). But it also requires the creation of a hydrogen-using infrastructure.

If zero-CO_2 by 2050 is defined as being within 5 percent of present CO_2 emissions, about 20 percent of electricity generation could come from natural gas combined cycle plants in that year. This would be a more than sufficient contingency for the failure of one or more of the advanced technologies that are part of the electricity sector in the reference scenario (V2G, hot rock geothermal, and biomass derived methane all put together, for instance). Further, the CO_2 from combined cycle plants can be captured and sequestered. As discussed in Chapter 3, carbon sequestration technology needs to be developed in any case as a prudent measure in case we need to recover some of the already emitted CO_2 from the atmosphere. Finally, sequestering 250 to 300 million metric tons of CO_2 would be qualitatively less problematic than attempting to find sound locations for disposal of amounts that would be several times larger, were coal to continue as a major energy source.

If such a contingency were to be put into action, alternatives for the remaining natural gas would have to be researched, developed and put into place for a complete elimination of fossil fuels. As discussed, such alternatives do exist, but it is difficult to estimate their commercialization prospects at present. Given that, it is possible that even with a vigorous and ongoing program of evaluation, research, development, and demonstration, achieving zero-CO_2 emissions in the literal sense could take a decade or so longer than in the reference scenario.

For coal to remain as a contingency in an economy with zero-CO_2 emissions, it will be essential to first demonstrate that carbon sequestration is a reliable technology that will contain CO_2 underground for thousands of years. The specific geologic settings and circumstances in which such performance can be expected will have to be specified. As noted in Chapter 3, the development of carbon sequestration technology is important in any case as a contingency in case the extraction of CO_2 already emitted to the atmosphere is needed. Such an eventuality may arise if climate change is far more severe than now anticipated in models that call for a 50 to 85 percent reduction in CO_2 emissions by 2050.

Some effort at developing approaches for removing CO_2 from the atmosphere at modest energy cost is also warranted. However, we note that resorting to this will increase energy use and complicate and possibly lengthen the schedule for eliminating CO_2 emissions.

In sum, natural gas combined cycle could be used as a contingency source of electricity power supply for up to 20 percent of generation in the reference scenario even if sequestration does not prove to be viable. For coal to serve as a contingency fuel in a zero-CO_2 economy, a prior demonstration that carbon sequestration would be feasible is necessary.

F. Structural Changes in the Economy

It is environmentally desirable to have many other changes in the structure of the U.S. economy that have not been factored into the reference scenario or any of the technical alternatives considered in this chapter. These do not relate to energy conservation as such, but rather to broader decisions about the pattern of economic development that could have significant implications for energy demand, and for the pace and the nature of the transition to a renewable energy economy.

For instance, in Chapter 4, we considered the issue of public transportation and showed that the energy efficiency and overall energy use in personal transportation (including cars) and even the number of cars owned varies according to the quality of the public transportation infrastructure. Were high quality public transportation to be treated as a public utility, a necessity for cities, much like electricity supply from a grid or sewage treatment systems or public water supply, the structure of cities would tend more toward being like San Francisco or New York or London or Paris. The mix of walking, public transport, bicycling, and automobile use would change, not because of energy considerations, but because it was more convenient and healthy, as well as less polluting. There is no evidence that such changes would decrease wealth or the GDP, but they would shift it toward greater public infrastructure investments and less energy production and consumption investments. The structure of the energy investments would also be different.

As another example, there are many reasons to consider greatly reducing the use of water sold in plastic bottles. Some leading brands of bottled water are just treated tap water. Transport of water over long distances contributes to water and air pollution needlessly. Despite recycling efforts, most plastic bottles are discarded. Finally, there is the question of the use of petroleum to make the plastic.

Much tap water, like that in New York City, is famously pure. Pollutants can be removed from tap water with commercially available filters at a small fraction of the cost of bottled water. A significant reduction of bottled water use would have modest implications for energy, but were it accompanied by similar changes in food and beverage consumption patterns, the implications for energy demand in the agricultural and industrial sectors could be significant.

One would not advocate a change were bottled water essential to health. But, arguably, it is not, as a general matter. Similarly, changes in where we live, what modes of transportation are available to us, and what we choose to eat and drink can have important effects on the shape of a renewable energy economy. This report shows that they are not essential to achieving it. However, a change towards a less energy intensive economic structure, because it is healthier and more desirable for other reasons, could accelerate the transition to a renewable energy economy much in the same way as increasing the efficiency of energy use. This topic is vast and complex in its own right. Moreover, it is not essential to the core investigation as to whether a zero-CO_2 emissions economy without nuclear power is feasible; hence, we have not attempted to quantify the effects of structural changes in the patterns of production, trade, and consumption. However, this omission should not be construed as an indication of a lack of importance of structural changes that improve quality of life, health, and reduce energy use.

G. Some Considerations in Setting Target Dates for Zero-CO_2 Emissions

We selected 2050 as a reference date for a zero-CO_2 economy for several reasons:

- The amount of installed coal and nuclear electric capacity in the United States is very large and it will take time to phase it out.
- It will be difficult to substitute liquid and gaseous fuels in the residential, commercial, and industrial sectors quickly and it will become more economical as equipment depreciates, new buildings are built, and existing buildings are sold.
- A number of the technologies that are needed are not yet fully commercial and some have not been fully demonstrated (such as using V2G to enable efficient use of renewable resources).
- The sunk investments in the fossil fuel sector would be largely lost if the equipment is retired prematurely.
- Rapid increases in the price of CO_2 allowances, for instance, by sharp reduction in CO_2 caps for the industrial and electricity production sectors, may cause a large-scale migration of industry offshore. Though this study has been done only in the U.S. context, it is recognized that there are limitations to actions in one country alone in terms of implications for global CO_2 emissions.

This is a powerful set of reasons. But at least as powerful is the quickly developing climate crisis, whose presence is clear but whose dimensions are as yet emerging. Adverse changes are occurring much faster than estimated even a few years ago. Hence the case for more rapid action is persuasive, at least to this author.

1. Historical Examples

Large transitions in the energy sector are nothing new. There was a huge transition from wood and animal power to coal in the nineteenth century. But it was still partial. Animals were still the main source of energy on farms, and the use of wood was still widespread a hundred years ago. Three other examples are more indicative of the potential for rapid transitions.

We have already discussed the first, which occurred in the United States after 1973. Within a couple of years, a relationship of lock-step growth between the economy and energy use that had been considered almost a law of modern economic development was broken. For over a decade, economic growth occurred without energy growth (on average). Industrial growth continued after that without energy growth. Hence, it appears possible to move the economy in a direction of more efficient energy use in a very short time. In this example it took an external shock. But there is no inherent reason why policies related to climate change cannot propel a similar change. We have already taken this into account to some extent in our demand scenario, though there is still ample room beyond that for energy efficiency, as noted above.

A more rapid transition can also be achieved if there are breakthroughs in supply and conversion technology. Let us first briefly consider some recent historical examples of major energy transitions.

The energy economy of the United States was thoroughly transformed in the first four decades of the twentieth century from horses and coal-fired trains to electricity and oil-fueled cars and tractors. The evolution since World War II has been of growth, not of structure. Nuclear power has not changed this fundamentally, since it supplies only about eight percent of U.S. energy and about 20 percent of U.S. electricity. Seen in this context, a time scale of about forty years appears to be reasonable and practical. The evolution of the energy economy was driven by a mix of laissez-faire, government policy, cheap oil, and two world wars.

The transitions in the electricity sector in France since World War II are even more interesting. There were two major ones between about 1960 and the 1990s. Table 6-1, taken from an earlier IEER report on the French energy sector, summarizes those transitions.

Table 6-1: French electricity sector transitions energy supply, in percent

	1960	1973	1996	2001
Coal [1]	~35	16	5	6
Hydro	56	27	14	14
Oil	7	39	Included in "other thermal"	2
Other thermal [1]	2?	10	3.6	1.4
Nuclear fuels	Negligible[2]	8	77	76

[1] We do not have exact coal and "other thermal" data for 1960.
[2] The initial sources of nuclear electricity in France were the plutonium production reactors in the nuclear weapons sector.
Source: Based on Makhijani and Makhijani 2006 Table IV.1 (page 27)

In 1960, the French electricity sector was dominated by hydropower and coal – they were over 90 percent of the total supply. In an era of rapid electricity growth and cheap oil France made a major electricity sector transition in only 13 years. By 1973, coal was on its way out, hydropower made half the percentage contribution it did in 1960, and oil had risen from 7 percent to 39 percent. Natural gas went from essentially zero to nearly ten percent. These rapid changes should be seen both as a result of national policy (France's electricity sector was

100 percent nationally owned) and rapid growth. Hydropower output did not decline. Rather, electricity use grew – and the growth was taken up by cheap oil and natural gas.

The second transition was more complete. Essentially the entire electricity sector, except for a more or less constant total contribution from hydro (and hence declining share), was supplied by nuclear power. This was not the result primarily of economics. France could have imported coal from the United States, for instance. It was the result of an "energy independence" decision taken to reduce France's dependence on petroleum, since almost its entire supply was (and continues to be) imported. While France is still dependent almost entirely on oil imports for its transportation sector, oil was nearly eliminated from the electricity sector.[13]

France's electricity sector transition shows that a nearly complete transition in a large sector can occur in less than 25 years, given determined government policy. It must be noted here that there was precious little consultation with the public on the transition to nuclear power, which has created its own problems, for instance, in terms of finding a site for disposal of high-level nuclear waste. The French government also owned the sector it transformed. But we see no fundamental reason why, with the right policies and incentives in place, a transformation of the U.S. energy economy to one that has very low CO_2 emissions cannot be achieved in 30 years, that is, before 2040. The French example shows that a transformation to a proliferation prone and costly technology that did not even solve the oil import problem in France was possible in the name of energy independence. The same could surely be done in making the change to an efficient, renewable energy economy given that, according to the Stern Review, climate change represents "the greatest and widest-ranging market failure ever seen."[14] The uncertainties largely lie, perhaps, in the last 10 or 15 percent of energy supply requirements.

2. Demand Sector Considerations for a Target Phase-out Date

Two complementary approaches to energy supply and to CO_2 emissions reduction could greatly accelerate the process. First, the residential and commercial sector should be considered together in terms of policy for encouraging renewable energy sources. The scale of residential solar PV is so small that custom backfitting will likely continue to be expensive even with cheaper solar cells, since the balance of the costs, including retail price markups, costs of inverters, connections to the grid, and labor, would not diminish very much. By contrast, medium-scale commercial installations in parking lots and on roof tops – 100 kW to a few MW – can be envisioned in the coming years at installed costs as low as $1.25 per peak watt. At the present time, the cheapest solar cell manufacturing is $1.25 per peak watt and installation costs are in addition to that.

As noted, bringing installed costs down to such a low level requires process improvements for solar cells that are already in the manufacturing and commercialization stages. But it does not require fundamental new technical breakthroughs. Broadening the concept of "zero net energy" could help. The term is usually defined in the context of a single building; it is taken to mean that the energy produced within the premises of the building (including its grounds and structure) is, on an annual average basis, equal to the energy consumed. On a day-to-day basis, energy may be imported or exported from the building, usually from and to the electric grid (respectively).

Contracts to sell electricity from commercial-scale installations to private residences and to other buildings in the commercial sector itself could be included in a community concept of zero net energy. This "community zero net energy" or "area zero net energy" could accelerate the transition to renewables by allowing development of lower cost resources first and making them available to a larger population.

There are already examples of institutional arrangements for contracts between commercial institutions. For instance, specialized companies are installing medium-scale solar PV on roofs and parking lots and selling the electricity to the corporations that own the buildings at their existing cost of electricity. The solar energy companies themselves make money from the electricity sales revenues and state, local, and federal rebates and incentives for solar PV.

Parking lot and rooftop area in the commercial sector is sufficient to supply both the residential and commercial sectors.[15] We estimate that, with time-of-use pricing, such contracts would not require incentives at $2 per peak watt or less (installed). One important constraint could be the quality of local distribution systems, which would need to be improved in many cases. Transmission costs are avoided. Equally important, if intermediate-scale systems form a principal source of supply, then the need for new transmission corridors can be reduced, and, in some cases, eliminated.

Local storage of electricity could also make the transition more rapid. As noted above, V2G systems and/or stationary electricity storage would allow a higher fraction of installed renewable capacity at the local level without placing large demands on the grid for providing reserve capacity. Either V2G or storage technologies are critical. If both can be successfully developed in a decade, the CO_2 emissions due to personal vehicles and residential and commercial electricity consumption, about 45 percent of the total, could be eliminated in about 30 years, possibly less.[16]

Michael Winkler has proposed an integrated electricity and thermal storage system. That storage can be accomplished using hot water. Storage of cold is accomplished with a specially designed ice-maker. Such a system could reduce the costs of a renewable energy system by minimizing installed capacity requirements. For instance, night-time wind-generated electricity could be used to make ice which would provide cool air during the day-time.[17]

H. Estimating a Phase-out Schedule

We will first consider a recent historical parallel to the complete elimination of a class of industrial materials due to environmental concerns and then summarize the possible range of dates by which CO_2 emissions could be eliminated without the use of nuclear power.

1. Ozone-depleting Chlorofluorocarbons (CFCs)[18]

The history of the complete elimination of CFCs, which were almost as ubiquitous as fossil fuels, though in more subtle ways, is instructive. In the mid-1970s, CFCs were used in everything from refrigerators and car air conditioners to the foam used for flower arrangements and insulation to solvents for cleaning electronic circuit boards to spray cans. In the 1970s, in a bow to initial scientific concern and findings and popular sentiment, the use of CFCs in aerosol spray cans was banned in the United States. There was as yet no detected large-scale depletion of the ozone layer.

In 1985, the existence of the Antarctic Ozone Hole was confirmed. By 1987, other trends in ozone layer depletion also showed themselves to be worse than previously estimated. In 1985, only the Vienna Convention for the Protection of the Ozone Layer was in place as an international treaty. It asked its members to take action to protect the ozone layer, but placed no numerical limits on emissions and had no phase-out date for CFCs.[19] There was widespread sentiment in the two to three years that followed for a complete phase-out of CFCs, but there was also much industry resistance and alarms about potential drastic economic and social results if CFCs were phased out.

Yet alternatives were available or nearly so. A report done by the present author with two other colleagues showed that alternatives existed in every sector where significant amounts of CFCs were used.[20] Some were not as economical as CFCs but others turned out to be cheaper. Some were in the pilot plant stage. Some were well developed. By 1987, when the Montreal Protocol to protect the ozone layer was signed, there was agreement to reduce CFCs production by 50 percent by 1998. But the crisis clearly demanded more. In 1988, DuPont, the largest manufacturer of CFCs, announced it would stop making them by the year 2000. In the same year, Sweden announced it would phase out CFCs by January 1, 1995. The 1990 revision of the Montreal Protocol, signed in London, set the year 2000 as the target date for a complete phase-out of CFCs by the developed countries. At the Copenhagen meeting of the parties to the treaty in 1992, the CFCs complete phase-out date was moved up to 1996. It was achieved. The developing countries were given an extra ten years.

The phase-out of CFCs was not without its bumps and problems. Some of the substitute compounds also caused depletion of the ozone layer, though not as powerfully. Some were greenhouse gases. At least some of these problems could have been avoided by a more thoroughgoing early elimination of ozone-depleting compounds than was agreed.[21]

The situation in the energy sector is similar, not only as a broad analogy but also in many details, such as the various stages of the development of the required technologies, the conflicts between partial reduction of CO_2 emissions versus a complete or near-complete elimination. Further, there are multiple goals to be achieved – in climate change, foreign policy as it relates to oil imports, and nuclear non-proliferation. A bold approach to eliminate CO_2 emissions, adopted early, with frequent and careful reconsideration of the potential for accelerating the schedule and also taking into account unanticipated problems, is indicated by the experience with ozone layer protection.

2. A Range of Dates for Zero-CO₂ Emissions

The energy sector is far larger and more complex than the use of CFCs. It will take investments and changes on a longer time frame, if only because the stock of existing capital – buildings, vehicles, aircraft, and industrial equipment – is so much larger. The main lesson of the rapid CFC phase-out was that with a firm target date that all parties knew would be enforced, CFCs were actually rapidly phased out at modest cost and little economic dislocation.

As noted above, there is no real technical obstacle to an elimination of the CO_2 emissions associated with personal vehicles and the residential and commercial sector within about 30 years. (We assume a starting date for serious action by 2010, since the enactment of legislation and the promulgation of regulations is likely to take about two years). By extension, it should also be possible to significantly reduce the use of petroleum across a broader swath of the transportation sector in that time. If the distributed generation of hydrogen and its use in internal combustion engines is put on the front burner of technology and infrastructure development, the whole land-based transportation sector could end petroleum use and move to a combination of electricity, hydrogen, and liquid biofuels. Each might be used alone, or two might be used in combination, as with plug-in hybrids for electricity and liquid biofuels, or dual-fuel internal combustion engines that use hydrogen and biofuels. If hydrogen can be economically compressed to 10,000 psi or more, it would be possible to have vehicles with reasonable range running only on hydrogen or a combination of hydrogen and electricity.

While hydrogen-fueled aircraft have been demonstrated, it is unlikely to contribute to a faster elimination of petroleum from that sector. The development of biofuels that resemble the properties of kerosene is more important for the air transportation sector.

One of the principal issues associated with biofuels is the amount of land that is likely to be needed for a complete transformation to a renewable energy economy if direct production hydrogen from solar energy is not developed and electrolytic hydrogen from wind energy is not made more economical. This throws some light on the importance of the development of the corresponding technologies for a more rapid phase-out. [22]

As discussed, all of the difficulties associated with the transition to renewable energy become more manageable if the efficiency of its use is increased to maximum feasible extent.

In sum, an elimination of fossil fuel use and nuclear power by about 2040 seems feasible if most of the following technical conditions can be met (policies are discussed in Chapter 7):

1. V2G technology is developed rapidly and/or stationary technology for electricity storage is developed rapidly so as to come down in cost to $200 per kWh or less. The main aim would be to make intermediate-scale solar PV supply most or all community electricity requirements. Investment in strengthening distribution systems would likely be required in some or many areas.
2. Greater use is made of solar thermal technology for heating and cooling in the residential, commercial, and industrial sectors, as well as for process heat in the industrial sector.
3. Efficiency is increased over that projected in the reference scenario, using technologies that are available today, along with greater electrification in the residential, commercial, and transportation sectors.
4. Wind-generated electricity is used to produce hydrogen on a large-scale, possibly using existing transmission corridors for creating a pipeline infrastructure. Alternatively, offshore wind development could be coupled with onshore distributed hydrogen infrastructure.
5. Greater use is made of hydrogen produced from wind energy in industry to produce feedstocks.
6. Direct solar production of hydrogen becomes economical within the next 15 years at efficiencies of ~10 percent, especially if such production can occur on an intermediate-scale, sufficient to serve single large factories or a few thousand automobiles. This allows faster incorporation of a significant amount of hydrogen into the fuel mix in place of liquid or gaseous biofuels.

The last item is, at present, in the stage of research. The other items in the list involve technologies that are already known and economical under some circumstances, or are within a factor of five of becoming economical. This last applies to ultracapacitors for large-scale stationary electricity storage and to lithium-ion batteries for electric vehicles. The cost of electrolytic hydrogen production is currently about a factor of two higher than the cost of gasoline, without taking into account any of the external health and security costs associated with oil.

The above list is not meant to be exhaustive of the possibilities that could result in an earlier elimination of CO_2 emissions from the U.S. energy sector. Rather, it is envisioned that a regular process of evaluation will take place to gauge the effectiveness of the policies, to assess new technologies, and to consider unanticipated problems.

As a final note on the feasibility of creating an efficient economy based on renewable fuels by about 2040, we note that the depreciation of most of the energy production, conversion, and utilization equipment occurs over the 10-to-40 year range. A modest acceleration of this, induced by a price paid for CO_2 emissions allowances, could produce a more rapid replacement of existing infrastructure, provided the technologies were available at reasonable cost. This puts a significant burden on government to get its policies right, to have a system for making mid-course corrections, and to shape the market by performance-based procurement policies that will enable needed technologies to be commercialized faster.

Prolonged difficulties, for instance, in commercializing liquid biofuels from high productivity biomass or failure to achieve significant cost reductions in lithium-ion batteries, would make some of the technologies not now in the reference scenario necessary for a zero-CO_2 emissions economy. Greater use of other technologies such as thermal storage for large-scale solar thermal power plants and solar heating would also be necessary. In turn, such a turn of events would tend to focus on power development in the Southwest where the number of sunny days is high. This would raise transmission issues.

One important contingency plan to prevent delays beyond 2050 is to maintain a significant portion of the natural gas combined cycle infrastructure for generating electricity. This would provide a margin for error and failure in other areas that could help prevent a slippage of the 2050 target. As noted, if natural gas combined cycle were used for 20 percent of the electricity generation in the reference scenario, the total CO_2 emissions would be less than five percent of the level in 2004.

Carbon sequestration technology would provide some redundancy, but it could be limited if there are significant problems in finding geologic sites for reliable, long-term disposal of CO_2. Finally, vigorous development of solar hydrogen production and development of hydrogen-fueled aircraft would also provide redundancy in case of problems with large-scale hydrocarbon biofuels production.

CHAPTER 7: POLICY CONSIDERATIONS

The atmosphere, and specifically, its role in regulating the Earth's climate has been treated with disregard – or in economists' parlance, it has been treated as a "free good." This disregard creates many problems, including market decisions not to make investments in reducing CO_2 emissions. In the absence of economic incentives or penalties for reducing emissions, incurring expenses to reduce emissions puts the good environmental actor at a disadvantage in the marketplace under many circumstances. But the problem goes far beyond that. For instance, when energy is a modest or small part of a company's or individual's budget, they may pay little attention to opportunities to save money even at existing energy prices. For instance, it is economical to change from incandescent to compact fluorescent lamps, but the former still continue to dominate the lighting market. Corporations have been more responsive to opportunities to reduce energy consumption because saving energy often increases profits. In the residential and commercial sectors, the market failure is structural. Developers of residential and commercial properties generally do not pay the energy bills, so that there is actually a built-in incentive to skimp on items that are not uppermost in the buyers' or renters' minds, such as energy efficiency investments. In this case, there are actually built-in incentives for inefficiency (the technical term is "split incentive").

A number of approaches can, in theory, be used to reduce and eliminate CO_2 emissions:

1. Fossil fuels can be taxed according to their carbon content.
2. Emissions of CO_2 can be taxed.
3. A cap can be placed on CO_2 emissions, with the total amount being periodically reduced so as to ensure that emissions are declining with time. This system was first introduced on a large-scale as part of the 1990 Clean Air Act for reducing power plant emissions of sulfur dioxide (SO_2).[1]
4. A cap can be put on total production and import of fossil fuels, with a total ban going into effect in a pre-designated year.

5. Certain uses of fossil fuels could be banned. For instance, there have been proposals to ban new coal-fired power plants.
6. Indirect methods, such as efficiency standards for buildings, appliances, and vehicles can be used to reduce the total amount of energy needed for a given level of economic activity.

These methods are not mutually exclusive. For instance, at the present time, the United States has both gasoline taxes and fuel efficiency standards, though both are quite low. The European Union has high gasoline taxes as well as manufacturers' agreement to meet efficiency targets.[2] As another example, the problem of ozone-depleting chlorofluorocarbon emissions was addressed by simply banning production of CFCs and importation into developed countries by a certain date (1995) and in developing countries ten years later. And appliance standards without significant electricity taxes have helped greatly reduce electricity consumption for the same levels of air-conditioning, refrigeration, etc.

Some economists prefer taxes as the most efficient way of internalizing the costs of pollution and hence, reducing it. If the level of tax is not high enough to achieve the goal, it can be increased until alternative fuels and efficiency become sufficiently economical to do the job. However, taxes would pose significant problems for large portions of the energy sector of the United States, notably in the personal transportation sector. The level of taxes needed to reduce gasoline consumption significantly is quite high, since gasoline is typically only about one-fourth or one-fifth of the operating expense of a personal vehicle (unlike, say, a taxi). In Europe, where gasoline taxes run to several dollars per gallon, the efficiency of cars is still far below what it could be with available technology. In the United States, gasoline prices have doubled in the past few years, without a significant reduction in demand. In the economists' jargon, gasoline demand for personal vehicles is rather inelastic – that is, its sensitivity to price is rather low in practice (though its political sensitivity is higher). Second, low-income people tend to have the oldest and most inefficient vehicles; that makes a high gasoline tax (or tax on petroleum) very regressive. In theory, the income derived from a tax could be redistributed to low-income households, but this redistribution would be complex and difficult to achieve in a fair manner, even if it were politically possible to actually put an adequate redistributive law in place. Third, a tax on one fossil fuel alone would distort the energy marketplace. For instance, a tax of petroleum would encourage investment in technology for turning coal into liquid fuels. A tax on vehicles that fall significantly below specified efficiency standards may be an effective complement to CAFE standards. The revenues could be used to provide incentives for vehicles with efficiencies far higher than the CAFE standards.

There is a better case for a carbon tax on all fossil fuels – it would be set according to the amount of carbon dioxide that would be emitted per million Btu of energy derived from burning that fuel. However, the level of this tax would have to be very high in order to affect the use of petroleum. A tax of one hundred dollars per metric ton of CO_2 corresponds to a less than one dollar per gallon of gasoline. While people would buy more efficient cars, the European experience makes it clear that it would be not adequate to reduce gasoline consumption sufficient to address global warming concerns. Yet a tax of $100 per metric ton of CO_2 is greatly in excess of what is needed for reducing and even eliminating CO_2 emissions from the electricity and buildings sectors. A carbon tax is a rather indiscriminate instrument that does not take into account the varying costs of reducing CO_2 emissions in different sectors of the economy. However, taxes may have a limited role in some circumstances as noted above.

We focus on the following policies as the main instruments for achieving a zero-CO_2 economy without nuclear power in the United States:

1. A combined fixed limit on CO_2 emissions per year for large fossil fuel users that would decline to zero in 30 to 50 years and sale of emissions allowances by the government corresponding to each annual cap.
2. Efficiency standards for vehicles, residential and commercial buildings, and appliances.
3. A shaping of the energy supply and demand marketplace through government procurement, research, development and demonstration, as well as preferences for government contracts to corporations that have relatively low CO_2 emissions for their sectors compared to prevailing norms.
4. Appropriate electricity rate structures at the state and local level.
5. A ban on new coal-fired power plants without CO_2 storage.
6. Elimination of subsidies for fossil fuels and nuclear power.

A. A CO_2 Emissions Cap Declining to Zero

The first large-scale implementation of a cap on emissions of a pollutant that would decline over time was for sulfur dioxide. It was enacted into law in the 1990 Clean Air Act. It applied to large electric power plants and then to power plants over 25 megawatts. Free emissions allowances were allocated to power plants in operation before 1995. Power plants that came on line in 1996 and after had to purchase allowances on the market or from the government. Trading in allowances is permitted. The Environmental Protection Agency administers the program. Any registered individual or institution can purchase or sell allowances. The cap is tightened periodically (in 2000 and 2010).[3] The program is important for the lessons it holds for CO_2 emissions. Its success in reducing SO_2 emissions in the United States made it a model for the European Union's CO_2 cap and trade program.

The European experience in CO_2 caps is the most extensive so far. The program is similar to the U.S. SO_2 program in that it applies only to large users, but it covers many more types of emitters than just power plants. The definition of large energy users adopted in the EU was quite complex, because caps were set for individual sectors:

> Large point sources were defined as power plants with thermal capacity of greater than 300MW, all refineries, sulphuric acid production plants and nitric acid production plants, iron and steel plants producing more than three million tonnes per year, pulp and paper plants producing more than one million tonnes per year, vehicle painting units painting more than one million vehicles per year, airports with greater than one million LTO [landing and takeoff] cycles per year, and any other activity producing more than one thousand tonnes of SO_2, NOx or NMVOCs [non-methane volatile organic compounds] or three million tonnes of CO_2 per year.[4]

Like the U.S. SO_2 program, free emissions allowances were granted to existing emitters of CO_2. However, since the varieties of emitters was much more complex, the problem of allocating emissions also was correspondingly complex. Further, giving free allowances based on prevailing use of fossil fuels tended to reward the most inefficient, since they got larger amounts of a marketable commodity, CO_2 emissions allowances, compared to more efficient companies.

Analyses of early results indicate that, in terms of reducing CO_2 emissions, it fell far short of what was anticipated. A study by the Öko-Institut of Germany examined the system, known as the European Union Emissions Trading Scheme (ETS), in some detail. Some of its main conclusions were:

1. Auctioning remains the most efficient allocation approach. All approaches based on free allocation of allowances to existing or new installations will face major problems in ensuring comprehensive and non-distorting incentive structures of the ETS (i.e. the full and comprehensive pricing of carbon). No Member State was successful in sufficiently balancing all different incentives (for existing installations, new entrants, plant closure and replacement) against each other, although some (e.g. the UK) did much better than others.
   ...

3. The criterion of economic efficiency should be seen as the most important especially with regard to existing installations in the power sector. Fairness problems mostly arise for the allocation to new entrants.
   ...

6. The full costs of carbon create the key incentive for the operation of existing power plants and the implementation of emission abatement measures in existing plants. Ex-post adjustments eliminate these incentives (see the German example).[5]

Many of the problems arose in relation to new entrants. With free and generous allowances for existing users, new entrants would be at a competitive disadvantage if they were not given free allowances. But if new entrants were given free allowances, the cap would have to be increased each time there was a new entrant into the market. Continual adjustments in the cap and issuance of allowances created a situation of an oversupply of CO_2 credits and a collapse of the market for CO_2.

A U.S. evaluation of the EU system concluded that for industries with large CO_2 emissions, caps at the point of fossil fuel combustion were effective. Such a cap is called a "downstream cap" because it is at the point of end use of the fuel, which is "downstream" of the fuel production, processing, and transportation system.

> A downstream system that focused on large energy users only would be more feasible [than one with universal coverage]. The number of regulated entities would be quite small: Per CORINAIR [European air pollutant emissions inventory] data, the number of large point sources in the fifteen EU Member States totaled only 1,652 in 1990. Further, the carbon embodied in fuel combusted would be easy to estimate based on existing fuel use records, and the regulated facilities would be experienced in reporting environmental data. Accordingly, much of the analysis in Section IV [of CCAP 1999] relates to a "limited" downstream system that covers large point sources only. [6]

For small users, imposing individual caps and attempting to enforce them would involve the creation of huge bureaucracies to administer the program. Small users number in the hundreds of millions. In 2004 there were about 230 million personal cars, SUVs, and light trucks and 113 million residences in the United States. In such a circumstance, some, including the Electric Power Research Institute, have advocated an "upstream" cap for small users:

> An upstream market-based system, one that requires fuel producers to surrender allowances or pay a tax for emissions attributable to their products would cover 90 percent or more of these emissions. [7]

According to this proposal, natural gas pipeline operators and/or natural gas producers would have emissions allowances and would pay a tax for continuing to sell natural gas to homes and businesses if they did not want to surrender their allowances. The same would apply to petroleum refiners who make gasoline and diesel. They would be paying a tax even though they do not actually use the fuel. Since they have huge investments in the existing fossil fuel infrastructure, they would have every incentive to pass on the costs.

On the other hand, developers, who make the basic decisions about the energy consumption structure of buildings, would receive only an indirect and weak signal regarding fossil fuel use, since they don't pay the energy bills. Moreover, for residential purchases, energy bills are generally a minor consideration in the purchase. Schools, safety, transportation infrastructure, and design features of the buildings are more central. And, as every real estate agent knows, the emotional factor – a house that a customer loves for its particular features – is often critical. Similarly, gasoline is only on the order of one-fourth the cost of operating a personal vehicle. As discussed above, taxes would have to be very high to have a significant effect on gasoline consumption. Moreover, there is no clear path to essentially eliminating CO_2 emissions, unless very high levels of taxation are imposed. A hybrid system proposed here would avoid the creation of a large bureaucracy while creating a framework within which almost all of the elimination of CO_2 emissions can occur (see Section B below).

Some lessons can be drawn from the European experience:

1. Free emissions allowances to existing users reward inefficiency, create inequities between new and existing users of fuels, and penalize those who have taken early action to reduce emissions.
2. Free allowances are relatively ineffective in reducing CO_2 emissions, especially in a context of trying to create a level playing field for new users of fossil fuels.
3. It is difficult to create a system covering all users when it comes to fossil fuels because of the very large number of consumers.
4. Auctioning allowances from the start is much more efficient than "grandfathering in" existing emitters and trying to add charges for new users only.

In light of the above, we propose the following policies to reduce and eliminate CO_2 emissions for large users:

1. A single "hard cap" – an absolute quantitative limit – would be set for all large users of fossil fuels together. It would be reduced every year and go to zero, by 2060 at the latest, with periodic evaluations to try to achieve it earlier. The term "users" includes electric utilities, since they burn the fuel in producing the commodity they sell. The definition of a "user" would be at the level of the holding company. The fossil fuel use of all subsidiaries would be added to determine whether the entity meets the definition of a "large" user.
2. The federal government would auction CO_2 emissions allowances to large users on a single open market, much like the sale of Treasury bills. However, in this case the number of CO_2 allowances would decrease each year until it reaches zero and the market would be national rather than global.
3. A penalty for fossil fuel use without allowances would be maintained at about ten times the average sale price of CO_2 allowances realized by the government in the prior year. This would discourage emissions without allowances.
4. Resale of unused allowances would be permitted.
5. Offsets would not be allowed – emissions would be allowed only against purchased allowances. In other words, fossil fuel users would not be permitted to emit CO_2 because they claimed that they have financed a reduction in emissions by a third party or planted trees somewhere that would supposedly capture the emitted CO_2.

This system incorporates market features in that it would allow holders of emissions allowances to use or sell them, since they have, after all, paid for them. But the more general "cap and trade" system that includes offsets and trading across borders would not be permitted. It is particularly important to avoid international offsets between countries that have set enforceable legal limits on emissions (whether by treaty or not) and those that have no such obligations. At present, CO_2 emissions' offsets purchased from developing countries create perverse incentives that could, and sometimes do, aggravate global warming problems. In the absence of a limit on CO_2 emissions, developing countries have an incentive to add to them if they can subsequently turn around and get paid for eliminating those same emissions.

It is not that the theory of offsets is without merit. Offsets, done within the framework of limits on emissions that are being tightened each year and enforced honestly, both within and across national boundaries, could produce more economical reductions in CO_2 emissions if they are measurable by strict criteria. *However, none of these basic conditions necessary for success is currently in place.* When an equitable and enforceable path to 50 to 85 percent reduction in global CO_2 emissions is worked out, offsets and international trading might be reconsidered. Until then, a national system without offsets is the surest way for the United States to proceed, especially as it is exceedingly well endowed with renewable energy resources and the opportunities for economical improvements in energy efficiency are great.

1. Early Action Rewards

A system of allowances in which all large users bid for them in a single market would also reward the companies that have invested early in CO_2 reductions as part of their corporate strategy, in anticipation of restrictions on emissions or as measures to save money or both. The United States Climate Action Partnership of corporations and private environmental organizations has made a particular point of the issue of providing appropriate recognition in practical, bottom line, terms to those who take early action:

> Prior to the effective date of mandatory emission limits, every reasonable effort should be made to reduce emissions. Those companies that take early action should be given appropriate credit or otherwise be rewarded for their early reductions in GHG emissions.[8]

An auction system would put those who take early action at a competitive advantage since they would have to purchase fewer CO_2 emission allowances. Another way that local, state, and federal governments could encourage action beyond the norm would be to award extra points, when evaluating government contract proposals, to those companies which excelled in performance on reducing CO_2 emissions. Companies could similarly adopt green purchasing policies; some already have such policies in place to varying extents.[9]

2. Defining "Large Users" of Fossil Fuels

Since it would be impractical, intrusive, and onerous to try to impose caps on small fossil fuel users, it is necessary to define the term "large user." We consider each of the two components of the term: "large" and "user."

Besides the practicality of enforcement, the term "large" must also be considered from the users' point of view. It would take some effort, experience, and expertise for a company to keep track of the CO_2 allowances market and determine whether it should invest to avoid emissions or purchase allowances for some more time. Such decisions would depend on the state of a company's finances and equipment at any given time and also on its view of its own future evolution. A company may decide to invest in energy efficiency after the purchase of allowances and sell the excess if the price of allowances goes up. The time and expertise invested in these decisions represent transactional costs of reducing CO_2 emissions, which should be kept well below the cost of the avoided fuel purchases.

A large user might be defined as one purchasing 100 billion Btu of fossil fuels or more. For an average future price of fossil fuels of $10 per million Btu, the threshold for fossil fuel expenditures would be $1 million per year. One hundred billion Btu is about equal to the delivered energy annually used by 1,000 households. A single 1,000 megawatt coal-fired power plant consumes about 700 times this threshold definition of large users. As another example, the definition would cover all large industries and corporations holding on the order of one million square feet of office space. It would also generally cover medium-scale industries and many small-scale industries. While there would be some paperwork requirements for all the entities defined as "large users," these would be kept to a minimum by having a single market for the CO_2 allowances and a single reporting time to the EPA each year (see below and also the interview with Dawn Rittenhouse and John Carberry of DuPont in Appendix B).

Electric utilities and independent merchant generators would be subject to the caps. Airline companies and large trucking companies would also be included in the caps. Fuel purchases for vehicle fleets owned by corporations would be included, but not personal vehicles owned by employees.

The term "user" would aggregate all the fossil fuel purchases of all subsidiaries of a corporation. Any other definition may encourage the formation of small subsidiaries that would each have fossil fuel purchases under the limit, giving such users an unfair advantage and also creating obstacles in reducing CO_2 emissions. In the commercial sector the definition would apply to the owners of the property.

The term "fossil fuel use" also needs definition. It is clear that it should include actual burning of fossil fuels because that is the activity that generates CO_2 emissions. Industries like oil refineries would be included only insofar as their own consumption of fuels was more than 100 billion Btu per year (which it generally is). Allowances would be needed only for the net amount of fuel they consume.

An important definitional problem relates to feedstocks. Over seven quadrillion Btu of fossil fuels, mainly petroleum and natural gas, are used as feedstocks for the production of a variety of goods, including basic chemicals, lubricating oil, pesticides, synthetic textiles and fibers, and plastics. These are not burned by the industries purchasing the fuels. However, much of the feedstock eventually degrade into CO_2, as for instance, when trash is burned in municipal incinerators. Keeping track of the fate of the materials made out of feedstocks could be even more onerous than creating caps for all users of fossil fuels. It is suggested therefore that feedstock uses of fossil fuels be included within the definition of fossil fuel "use." The use of fossil fuels in large amounts for feedstocks would fall under the cap, according to this definition.

In 2004, electric utilities and industries accounted for about 54 percent of total fossil fuel use. In addition a large portion of the transportation sector, such as airline companies, large trucking companies, and corporate vehicle fleets, as well as a significant part of the commercial sector, would fall in the large user category. An additional few percent would be represented by large truck and light vehicle fleets. However, not all commercial buildings or industries would fall under the term "large users." Overall, about two-thirds to three-fourths of total fossil fuel use would be covered by the cap. Residential sector purchases of fossil fuels, which consist mainly of natural gas and heating oil, and purchases of vehicular fuel for personal use and by small businesses would not be covered. But residential and commercial purchases of electricity from the grid would be affected by the cap so long as electric utilities are still using fossil fuels for electricity generation.

We reemphasize that the system is envisioned as a pure CO_2 permit system, with declining caps. Those who emit CO_2 would actually have to hold the allowances to do so, purchased at auction from the government or on the open market. CO_2 offsets, such as emitting CO_2 and claiming CO_2 capture in tree farms, etc., would not be permitted. The complexities of measurement of CO_2 balance in the soil, for instance, would create enforcement nightmares. Offset schemes tend to undermine the CO_2 market. Further, as noted, international offset schemes would face problems of huge loopholes and verification, notably in the absence of a binding global treaty with intra- and trans-national enforcement provisions to greatly reduce CO_2 emissions. Biofuels would be exempt from the cap. However, use of fossil fuels on a large-scale in producing biofuels would be included.

A modification of the system above can be considered to include emissions of greenhouse gases other than CO_2 that occur in the energy sector. For instance, there are emissions of methane associated with pipelines and emissions of certain other gases such as HFCs from industry. A total CO_2 equivalent cap corresponding to emissions from the covered entities (large users of fossil fuels) could be set. This would likely be more desirable since companies would have the flexibility to reduce those emissions that are the cheapest to eliminate first. But it should be done with some rigor – and with measurability, enforcement, and verification as key considerations.

3. Penalties

Enforcement of the CO_2 cap for large users requires that they face some penalties for emitting CO_2 without holding an allowance to do so. The successful enforcement of SO_2 caps provides a useful guide. A penalty of $2,000 per ton is imposed for emitting SO_2 without holding an allowance. The level of the penalty is much more than the cost of reducing SO_2 emissions:

The SO_2 program has also brought home the importance of monitoring and enforcement provisions. In 1990, environmental advocates insisted on continuous emissions monitoring, which helps build market confidence. The costs of such monitoring, however, are significant. On the enforcement side, the Act's stiff penalties – $2,000 per ton of excess emissions, a value more than 10 times that of marginal abatement costs – have provided sufficient incentive for the very high degree of compliance that has been achieved.[10]

The same approach can be used for CO_2. The costs of reducing CO_2 emissions are expected to range from negative up to perhaps $40 per metric ton. A reasonable starting value of penalty would be about $100 per metric ton of CO_2, since the typical cost of abatement of CO_2 emissions in the early stages would likely be on the order of $10 per metric ton. A policy to maintain the penalty at about ten times the average sale price of CO_2 emissions in the prior year would serve as an effective enforcement tool. It would be expected to increase from the initial value of $100 per metric ton to several hundred dollars per metric ton as the use of fossil fuels declines, the cap is reduced, and allowances become more expensive.

In the SO_2 reduction system, the EPA requires electric utilities (only utilities are covered) to submit both the emission allowances and emission measurements for the preceding year. This system allows companies to adjust their operations during the year. They can purchase additional allowances, sell some of the ones they hold, and/or install pollution control equipment to reduce them, according to their estimate of the profitability of these measures. A similar system can be put in place for fossil fuels. The allowances would correspond in this case for fossil fuel purchases unless the user can show measurements that CO_2 has been captured, resulting in avoided emissions.

4. Revenues

Important practical economic goals are served by auctioning all allowances and setting an initial cap that is stringent enough to yield a non-negligible price but not so high that it would cause large business dislocations in the short-term. For instance, if the auction price averaged $10 a metric ton of CO_2 emissions,[11] a cap covering large users' emissions of about 4 billion metric tons of CO_2 would result in a total revenue of $40 billion per year. Four billion metric tons corresponds to about two-thirds of CO_2 emissions in 2005. As the cap is reduced each year, the price of each allowance would tend to rise. While it is difficult to estimate revenues over the long-term from such a scheme, one might anticipate that revenues would remain in the $30 to $50 billion per year, provided technological breakthroughs do not reduce the cost of eliminating CO_2 well below current estimates (see Table 2-1, Chapter 2). Breakthroughs are to be desired of course, since they would reduce the time required for a transition to renewables. They would also reduce the scale of government expenditures and investments in research, development, and demonstration plants, as well as added procurement expenditures required to shape the market along more efficient, renewable lines.

If there are too many allowances on the market, it would depress the price of a CO_2 allowance that the federal government gets at auction. This would indicate that the there is a greater potential for reducing CO_2 emissions at a given cost than anticipated. A falling price could therefore be a signal to the federal government to reduce the allowances for sale in future years, thereby accelerating the transition to a zero-CO_2 economy.

B. Small Users of Fossil Fuels

As discussed above, the imposition of caps on small users is impractical and would create inequities. But small consumers must also be brought into the overall scheme, since the required reductions on CO_2 emissions cannot be achieved unless they are. It is important to take into account the fact that individuals and very small businesses simply do not have the wherewithal to assess energy and environmental questions on a day-to-day basis. Further, the individual's control of the market is weak, though collective consumer preferences, such as for types of vehicles and homes purchases, do have a profound effect. Further, as noted above, developers and manufacturers of appliances and vehicles are small enough in number that efficiency standards can be enforced. Finally, efficiency standards on new equipment and buildings solves the problem of the "split incentive"– that is, the lack of incentive on the part of developers to invest in efficiency beyond required codes since energy bills are paid by owners or renters.

Standards for appliances and new buildings are easier to conceive and implement than standards for existing buildings. There is ample precedent for incremental tightening of efficiency standards for new equipment. Limits on Btu of externally delivered energy per square foot can be made part of state and local building codes and incentives can be provided for exceeding the standards. This is a performance-based approach, which allows the builder to decide what mix of passive features (such as building thermal mass and insulation) and active features (such as solar water heating or solar PV) to use to meet the code's requirements. The added costs, if any, become part of the mortgage payment. This is also the simplest way to finance the transition in the building sector. Gradually a zero net energy goal can be created – that is, imports of energy into areas and communities (purchased fuels and electricity) would equal exports when averaged over two or three years.

Similarly, costs of vehicle efficiency improvements become part of the cost of the vehicle. Any added costs for more efficient vehicles would become part of loans, if they are taken, to finance cars. The added cost would be largely or fully offset by reduced energy costs.

For existing buildings, the time of application of standards would be when they are sold. That way, the financing of the changes becomes a part of the mortgage taken by the new owner. Since it is more difficult and expensive to improve the efficiency of existing homes, the standards of existing buildings would be tightened more gradually and remain less stringent than those for new buildings.

1.   Time-of-use Rates

We have discussed the importance of time-of-use (TOU) rates in the context of the economics of solar energy during peak hours. A transition to a renewable economy would be greatly aided by more general adoption of time-of-use rates, especially since it would encourage investment in small- and intermediate-scale solar PV systems. TOU rates require a change of metering arrangements, since special meters are needed to measure electricity use according to the time of day. Net metering is a natural complement to time-of-use rates, since it both charges consumers at the rate then prevalent and also gives the consumers the corresponding rate when they supply electricity back to the system.

The oil and natural gas peaking systems operating for a few hours a day are the most costly. If the natural gas systems are single-stage gas turbines, which have very low capital costs but high fuel requirements, peak electricity costs, delivered to residential customers, can be as high as 20 cents per kWh (for natural gas costs of $8 per million Btu and single stage turbine capacity use of 300 hours per year). Costs of oil-fired peak generation would be similar or higher.

A flat rate for electricity grossly distorts the actual costs incurred and cannot be justified on market-based considerations. Since solar energy provides most of its generation during peak hours (and the rest during intermediate load hours) time-of-use metering is an action that corrects a large market distortion and promotes solar PV at the same time.

In a distributed grid supplied mainly by solar and wind energy, lower rates may not necessarily be at night, as is the case at present. Rather, rates would be high at the time of lowest supply in relation to demand. Flexibility would be introduced into the system through electricity and thermal storage and possibly a "smart grid."

2. Incentives and Rebates

In the initial stages of development of renewable energy sources and the encouragement of their use, rebates and tax incentives have been critical to their rapid growth. The Western Governors' Association has a goal "30,000 MW of clean, diversified energy" of which 4,000 MW will be solar PV (3,000 of it in California alone). Half-a-million solar thermal systems are also planned.[12] California has provided high incentives to early adopters (Table 7-1). The incentives are expected to decline significantly as more and more capacity is added. For instance, the incentive payment per kWh for the third tranche (MW Step 3) is 34 cents per kWh, if the capacity is in the residential or commercial sector. For the tenth step, the corresponding payment is only 3 cents per kWh. Payments are higher if the capacity is added by non-profits or the government. The California Public Utilities Commission had extensive public hearings and consultation with producers, consumers, and manufacturers in arriving at these incentives. These were accompanied by extensive analysis.[13]

Table 7-1: California Payment Scheme for Solar PV

Levelized PBI Monthly Payment Amounts at 8% Discount Rate			
MW Step	MW in step	PBI payments (per kWh)	
		Residential/Commercial	Government/Non-Profit
1	50	n/a	n/a
2	70	$0.39	$0.50
3	100	$0.34	$0.46
4	130	$0.26	$0.37
5	170	$0.22	$0.32
6	230	$0.15	$0.26
7	300	$0.09	$0.19
8	400	$0.05	$0.15
9	500	$0.03	$0.12
10	650	$0.03	$0.10

Source: CPUC 2006 Table 5, (pages 37-38)
Notes: 1. PBI = Performance Based Incentives
2. The increments in capacity are divided into ten steps. Each increment represents a total addition to capacity. The additions in each step are larger than in the prior ones. The earlier steps get higher rebates than subsequent additions.

California plans to spend $2.5 billion to $3 billion in implementing its 3,000 MW "Million Solar Roofs" program. This will be paid for by a charge of about 0.1 cents per kWh on electricity over a ten year period. [14] California also has rebate programs for zero-emission vehicles, which are helping to establish an initial market for electric cars.

Rebate programs are also important for encouraging the use of technologies that are very efficient but are marginally economical due to high first cost, such as earth-source heat pumps. There are many examples of incentives in essentially every state. They include residential and commercial installations. [15]

3. Achieving Zero-CO_2 Emissions for Small Users

The policies discussed above would result in large reductions in CO_2 emissions by smaller users, but would not guarantee zero-CO_2 emissions. Some individuals may want to continue using fossil fuels. Further, most large users, as defined above, would fall into the small user category at some stage as they reduce their fossil fuel consumption. The absence of carbon taxes would create the potential for fossil fuel prices to decline below the prices of renewable fuels as large users become more efficient and switch to renewable fuels. Such a situation would likely not occur for a considerable time (at least two decades). But, in the long-term, supplementary policies may therefore be necessary to ensure a continued transition to a fully renewable energy economy, including

1. Zero-CO_2 emissions requirements for developers of new buildings.
2. Zero net energy goals for areas and communities (in combination with a grid consisting of renewable electricity only).
3. Emissions or fuel type requirements for new vehicles.
4. A ban on fossil fuel production and imports by a certain date, similar to the CFC ban.

It is possible that some combination of the first three policies would be required unless the fourth is used. [16]

C. Government Actions

A shaping of the energy supply and demand marketplace through government procurement, research, development and demonstration is part of the solution for achieving a more rapid transition to a zero-CO_2 emissions economy. Some of the estimated $30 billion to $50 billion in annual revenues derived from the sale of CO_2 emission could be transferred to state and local governments for supporting programs analogous to those in California and other states that have already taken the leadership in promoting efficiency and renewable energy sources.

Plug-in hybrids could become the standard issue government car by 2015.

Large-scale central station solar energy plants to stimulate investment in large-scale solar PV manufacturing and in solar thermal technology are needed. Lack of sufficient demand is the central obstacle that is preventing economies of scale from being achieved in critical technologies. Demonstration of V2G technology on a scale that would test its viability for creating a reliable grid is also needed. A more detailed list is specified in a timeline in Chapter 8. Taxi commissions in cities can allow (or require) taxis to be hybrid cars. Federal, state, and local governments could set zero net energy, or at least zero net electricity goals, to be achieved in about 20 to 25 years.

The federal, state, and local governments can also give preferences in contracts to corporations that have relatively low CO_2 emissions for their sectors compared to prevailing norms. Some corporations have already adopted such policies in their own purchasing decisions (see Appendix B).

One important initiative would require collaboration between the federal, state, and local governments. Aquatic plants can be grown in the effluent of waste water treatment systems, particularly if these are combined with constructed wetlands. There are a host of regulations that already cover wastewater treatment. Integrating biomass production with them would be a complex regulatory question. However, given that (i) plants like water hyacinths have been shown to improve water quality (see Chapter 3), and (ii) they have the potential to contribute significantly to energy supply, a joint exploration of the ways to accomplish that along with demonstration projects in various climates should be an important funding priority A demonstration of offshore wind energy, coupled with onshore electrolytic hydrogen production, is also desirable.

Finally, a fundamental change in the sources of energy supply in the U.S. economy will no doubt affect large numbers of workers, from coal mining and petroleum to suppliers of automobile parts. Fossil fuels are mainly produced today in the Appalachian region, in the Southwest and West and some parts of the Midwest and Rocky Mountain states. For the most part, these areas are also well-endowed with the main renewable energy resources – solar and wind. In the East and Southeast, offshore wind is a significant resource. Distributed hydrogen production and utilization infrastructure could be a major new industry. Federal, state, and regional policies, designed to help workers and communities transition to new industries, therefore appear to be possible without more major physical movement or disruption of populations than has occurred in post-World War II United States. It is recognized that much of that movement has been due to dislocation and shutdown of industries, which causes significant hardship to communities and workers. Some of the resources raised by the sale of CO_2 allowances should be devoted to reducing this disruption. For instance, the use of CO_2 capture technologies, notably microalgae CO_2 capture from existing fossil fuel plants, can create new industries and jobs in the very regions where the phase-out of fossil fuels would have the greatest negative economic impact. Public policy and direction of financial resources can help ensure that new energy sector jobs that pay well are created in those communities.

D. New Coal-fired Power Plants

New coal-fired power plants that do not have provisions for capture and sequestration of CO_2 should be prohibited. New pulverized coal-fired power plants would have a life of about 40 years or more. Since these plants are now quite expensive, the owners of new ones would constitute a formidable lobby to advocate slowing down, diluting, or stopping mandatory reductions in CO_2 emissions. Since wind-generated electricity is already economical relative to coal with sequestration, there is no reason to allow the building of new power plants that would emit large amounts of CO_2 for decades.

E. Ending Subsidies for Nuclear Power and Fossil Fuels

Nuclear power still gets a significant subsidy in the form of government-provided accident insurance. Further, despite all the talk of a nuclear power renaissance, not a single new nuclear power plant has been ordered as of this writing (July 2007), despite added subsidies for license application and other costs that were enacted into law as part of the Energy Policy Act of 2005. Congress is considering 80 to 100 percent loan guarantees for new power plants, that may extend to as many as 28 plants, at $4 billion to $5 billion each.[17] Even so, Standard & Poor's, the well-known Wall Street credit rating agency, has stated that:

> ...an electric utility with a nuclear exposure has weaker credit than one without and can expect to pay more on the margin for credit. Federal support of construction costs will do little to change that reality.[18]

This means that Wall Street, or at least an influential portion of it, considers nuclear power such a high risk that the credit rating of a utility ordering it would be likely to suffer, even if the federal government provides subsidies. The result of an order would, therefore, likely increase the costs of electricity across the board, making any utility that ordered a nuclear plant less competitive.

The escalating costs of finding, characterizing and developing a deep geologic repository program for nuclear waste provide an added element of risk. Expanding nuclear power plant capacity significantly will likely require a second repository, when it is already unclear whether the proposed Yucca Mountain repository for disposing of spent fuel can ever be licensed. The site's deficiencies have been extensively written about, including by the present author.[19] Adding more nuclear power plants risks more repositories, higher costs for repositories, or higher costs for reprocessing, or all three. Further, heat waves and droughts may cause nuclear power plants to be shutdown for extended periods at times of peak demand. Since such events are expected more frequently in a warming world, an element of intermittency may be introduced into nuclear energy.

Massive subsidies should not be sustained indefinitely for any source of energy, and especially not one that carries significant nuclear proliferation, waste, and severe accident risks. Nuclear power advocates claim that it could be part of the solution of the climate change problem. CO_2 emission caps will cause the costs of fossil-fuel-related generation to increase. Nuclear power should be able to compete with that in the marketplace. There is no sign that it will be able to do so. Nuclear power should be eliminated from the U.S. economy as the current plants reach the end of their licensed lives.[20] Specifically, the following policies should be adopted:

1. All subsidies for new nuclear power plants, including government-supplied and guaranteed insurance, tax credits, and licensing subsidies should be ended.
2. Government should explicitly declare that it will not take responsibility for nuclear waste disposal from new nuclear power plants and that its responsibility extends only to existing power plants for their licensed lifetimes.
3. A regulatory infrastructure for reactor safety for existing reactors and for waste management and disposal should be maintained.
4. Onsite storage of spent fuel should be hardened against terrorist attack.
5. The insurance provisions for present plants should more realistically reflect the estimated damages from worst-case accidents that are estimated to be part of the plants' design vulnerabilities.
6. The ban on reprocessing spent fuel enacted under President Carter should be re-imposed.

Fossil fuels have been around far longer than nuclear power. Subsidies and tax breaks or loan guarantees for new applications, such as processing coal to produce liquid fuels, are especially counterproductive at a time when public policy needs to focus on achieving CO_2 emission reductions in ways that will not aggravate other problems. The exception that we would make to this policy is the full commercialization of IGCC technology, because essentially the same technology that is now proposed for coal would also be useful for electricity generation using biomass as a fuel. Carbon sequestration should also be developed for the reasons that have been discussed in Chapters 3 and 6.

F. Corporate and NGO Actions

The potential for a regulatory zero-CO_2 goal to achieve change is being illustrated in the marketplace, even from consideration of goals that are far short of this plan. For instance, the U.S. Climate Action Partnership (USCAP), which consists of corporations and large environmental non-government organizations, published a report advocating a U.S. target of 60 to 80 percent absolute reduction in greenhouse gas emissions by 2050.[21] This goal is reminiscent of major industries agreeing to the 1987 Montreal Protocol to protect the ozone layer, which required a 50 percent reduction of CFC emissions in about ten years. Eventually more was required, and developed countries phased out CFC production by 1996.

In February 2007, after the publication of USCAP's recommendations, a private group sought to complete the largest corporate buyout in history, that of TXU, which was planning to build 11 coal-fired power plants. The private group consulted with large environmental groups who were certain to oppose the deal. The cancellation of eight of the power plants and a plan to increase the building of renewable energy sources was the result.[22]

These actions, which have commanded a great deal of media attention, are only the most recent and most visible phase of a quieter but nonetheless important change that has been occurring. Insurance companies and some banking sectors of Wall Street have had practical concerns about global warming for some time. Multinational corporations that operate in scores of countries now have to deal with vastly differing rules in different places. Oil and gas companies face massive disruption in the case of more frequent and/or more severe loss of offshore production capability due to storms. Wild gyrations in natural gas prices like those that have occurred since 1999 make corporate planning much more difficult at higher levels of energy use. Turbulence in key oil and gas producing parts of the world has made planning for higher energy productivity a much higher priority in many boardrooms. A part of the result can be seen in the fact that energy use in the United Stated declined in 2006 to below the 2004 level.

Some corporations have been willing to be more open to outside advice and to analyses by nongovernmental organizations (NGOs), who may have been regarded not too long ago as adversaries. Tough negotiations were involved in achieving the cancellation of eight TXU coal-fired power plants. But the remarkable thing is that they took place at all and achieved a significant result.

CHAPTER 8: ROADMAP FOR A ZERO-CO$_2$ ECONOMY

It is technologically and economically feasible to phase out CO$_2$ emissions and nuclear power at the same time. The analysis in this report indicates that it can be done at reasonable cost by 2050. The goal could be achieved about one decade earlier, if biomass and hydrogen can be produced with high efficiency of solar energy capture and if greater efforts at energy efficiency are made. As discussed in Chapter 6, it is also possible that addressing some issues, such as creating a distributed grid with several new technologies, may take longer. The most important step at the present time to ensure the phase-out happens is to set a mandatory goal of a zero-CO$_2$ emissions U.S. economy as much before 2060 as possible. We first set forth a preferred renewable energy scenario to frame the detailed timeline. The action plan in the timeline also contains the contingency elements that provide redundancy in case the preferred approach cannot be realized to its fullest.

A. A Preferred Renewable Energy Scenario

Various possible components of an approach that would be preferable to the reference scenario were discussed in Chapter 6. This roadmap stresses a renewable energy economy based on a desired outcome rather than in the reference scenario. The main problem in the reference scenario is the relatively large area of land that would be required to cultivate the biomass needed mainly for liquid and gaseous biofuels that would replace fossil fuels in all sectors of the economy. Another problem is that the large amount of liquid and gaseous biofuels results in large energy losses. Five to six percent of the land area of the United States (and possibly more) would be needed. Impacts in particular regions would be considerably greater. While this is within the realm of feasibility, setting a course for a more efficient economy, with a component of hydrogen derived from wind and solar energy would be preferable.

Besides considerations of land area, there may also be issues of water use both in biomass crop production and in their processing into fuels. In view of these considerations, policy should seek to have considerably greater efficiency in all areas where liquid or gaseous biofuels are involved. The following appears to be a reasonable approach for that portion of energy demand relative to the reference scenario (electricity use and use of solid biomass for electricity generation remain unchanged):

- A significant reduction in use of gaseous biofuels in the residential and commercial sectors, for instance through greater efficiency and greater use of solar thermal heating. This applies mainly to space and water heating.
- A significant reduction in use of liquid biofuels in transportation through greater efficiency than in the reference scenario. As discussed in Chapter 6, the reference scenario assumptions are not very ambitious in relation to presently available and foreseeable technology.
- A reduction in biofuel requirements for feedstocks and fuel uses in industry though greater efficiency and greater use of solar thermal energy.

Some of the remaining hydrocarbon biofuel demand could be met using hydrogen in industrial combustion engines, greater use of electricity in the residential, commercial, and transportation sectors, and in industry. We assume that aircraft, much industry and most long-distance road transport will still use liquid biofuel hydrocarbons.

If these technological goals were realized, the overall biomass requirements would be significantly reduced. Electricity production would increase somewhat. And there would be a role for hydrogen in transportation (probably in internal combustion engines) and a greater role for hydrogen in industry. Hydrogen would be produced by a combination of electrolysis using wind energy and by one or more direct solar hydrogen production methods. In this preferred scenario, the land requirements for biofuels could be reduced to 2 to 3 percent of the U.S. land area (compared to 5 to 6 percent in the reference scenario).

Realizing this preferred renewable energy scenario would require:

- More stringent standards for buildings and vehicles compared to the reference scenario.
- Extended adoption of the concept of zero net energy beyond buildings to areas, communities, and institutions.
- Greater emphasis on research, development, and demonstration of electrolytic hydrogen from wind energy.
- Full commercialization of at least one technology for direct hydrogen production from solar energy in the next twenty years.
- Ensuring through government procurement and other incentives that, once the hydrogen production and use technologies are close to commercializa-

tion, that the infrastructure for its use will be created. Distributed hydrogen infrastructure – that is, infrastructure close to the point of use can probably be realized more expeditiously than a centralized system.

B. Timeline for Transformation

The following is a brief timeline based on the analysis in this report. The list is not comprehensive but indicative and based on the technologies that appear to be important at this time.

2007

1. Enact a physical limit of CO_2 emissions for all large users of fossil fuels (a "hard cap") that steadily declines to zero prior to 2060, with the time schedule being assessed periodically for tightening according to climate, technological, and economic developments. The cap should be set at the level of some year prior to 2007, so that early implementers of CO_2 reductions benefit from the setting of the cap. Emission allowances would be sold by the U.S. government for use in the United States only. There would be no free allowances, no offsets, and no international sale or purchase of CO_2 allowances. The estimated revenues – approximately \$30 to \$50 billion per year – would be used for demonstration plants, research and development, and worker and community transition.
2. Eliminate all subsidies and tax breaks for fossil fuels and nuclear power (including guarantees for nuclear waste disposal from new power plants, loan guarantees, and subsidized insurance).
3. Ban new coal-fired power plants that do not have carbon storage.
4. Enact high efficiency standards for appliances at the federal level.
5. Enact stringent building efficiency standards at the state and local levels, with federal incentives to adopt them.
6. Enact stringent efficiency standards for vehicles and announce the intention of making plug-in hybrids the standard U.S. government vehicle by 2015.
7. Put in place regulations requiring the recycling of batteries used in plug-in hybrids and electric cars.[1]
8. Put in place federal contracting procedures to reward early adopters of CO_2 reductions.
9. Establish a standing committee on Energy and Climate under the U.S. Environmental Protection Agency's Science Advisory Board.

2008–2009

1. Publish draft regulations and their finalization for treating CO_2 as a pollutant, cap and trade, etc.
2. Publish and finalize governmental purchase rules for biofuels to include liquid fuels made from microalgae .
3. Begin government purchase of plug-in hybrids.

4. Increase funding for the National Renewable Energy Laboratory (NREL), including an acceleration of the solar hydrogen and electrolytic hydrogen program.
5. Commission an evaluation of programs and policies (such as rebates, rate structures, etc.) in California and other states for applicability across the country.
6. Create an NREL program to evaluate and develop the uses of aquatic plants as energy sources.
7. Create a joint federal-state-local government task force on growing biomass for energy on constructed wetlands and begin planning pilot and demonstration projects.
8. Fund the following in collaboration with industry:
    *   Design of Integrated Gas-Turbine Combined Cycle plant for biomass, especially for high productivity biomass.
    *   Research on and development of nanocapacitor (supercapacitor) storage.
    *   Large-scale demonstration plant for the production of liquid fuels and methane from microalgae.
9. Commission a thorough optimization for integrating wind and solar electricity with hydropower and combined cycle natural gas standby into a distributed electric grid. The study should also explore the concept of a "smart grid," which integrates electrical and thermal storage components.[2]
10. Commission an economic impact study for areas with high fossil fuel production to devise policies for a just transition to a renewable energy system.

Also in this period a number of actions would be needed to prepare for a first test of a vehicle-to-grid system. A V2G Task Force – a joint federal effort with Independent System Operators in cooperation with one state (such as California) where the institutional infrastructure is already in place – would be created to carry out and evaluate such a test.

2010–2020

1. Begin implementation of the hard cap for large fossil fuel users at about the 2005 level of CO_2 emissions. It would be set to decline by 3 percent per year relative to the base year in the first ten years, and adjusted thereafter.
2. Begin a policy of installing roof-top and parking lot solar PV installations at federal facilities with a goal of making the federal government buildings a zero-net energy institution by 2030 or 2035 and begin revenue sharing with the state and local governments for the same purpose.
3. Build and test 5,000- to 10,000-vehicle V2G systems in three different regions.
4. Build several demonstration plants, from small to large, for growing high productivity plants (microalgae, water hyacinths, duckweed, etc.), in conjunction with wastewater treatment plants or in areas where runoff that is

high in nutrients is creating ecological problems. Build at least one plant where wastewater is piped out of metropolitan areas to areas with degraded land for biomass and biofuels production.

5. Continue development of fuel cells, especially for stationary applications.
6. Construct an electrolytic hydrogen plant for testing and demonstrating infra-structure for hydrogen for internal combustion engine vehicles.
7. Begin building pilot plants for promising solar hydrogen technologies.
8. Begin and complete construction of a 1,000 MW solar thermal plant with twelve-hour energy storage.
9. Enact building standards at the state and local level for residential and commercial buildings.
10. Begin designing and building an IGCC plant using biomass with no coal or other fossil fuels.
11. Complete evaluation of liquid and gaseous fuel production from microalgae, prairie grasses.
12. Design and build a pilot plant for liquid and gaseous fuels from aquatic plants.
13. Design and build a demonstration plant for nighttime storage of carbon dioxide emitted from fossil fuel plants with the aim of using the CO_2 to grow microalgae in the daytime.
14. Begin using liquid fuels from microalgae on a commercial scale in the 2015 to 2020 period.
15. Design and build a demonstration hot rock geothermal plant.
16. Ensure that all housing subsidized by the federal government, including housing provided with government-subsidized loans or insurance, is built to at least Gold LEED standards. (LEED stands for Leadership in Energy and Environmental Design; it is a building certification program.)
17. Conduct a study evaluating the amounts by which public transit riders subsidize automobile users in high traffic cities.
18. Complete an evaluation of the wind farm with compressed energy storage planned for Iowa and commission second generation demonstrations.[3]
19. Build an offshore wind-energy-based electrolytic hydrogen demonstration plant for distributed onshore hydrogen production
20. Begin design and construction of demonstrations of CO_2 sequestration, with a research design that will allow evaluation of the risks of leaks and the potential for sudden releases of CO_2 after disposal.
21. Build a large-scale Fresnel lens solar concentrator solar photovoltaic power plant.
22. Evaluate and put in place a program for hydrogen-fueled commercial aircraft, including a demonstration project.
23. Issue biennial reports from the EPA's Energy and Climate Committee, which would allow updating of the program for eliminating CO_2 emissions.

Toward the end of this period, the backbone of the energy system is transformed. At this stage, about half of the electricity and half of the total energy inputs would come from renewable sources. Major changes in the efficiency of the U.S. economy will have become institutionalized. Different ways of doing business will have become the norm. The CO_2 cap will have declined to about half of the base level in the 2025-2030 period, possibly lower. A mix of storage technologies, solar thermal power stations, solar PV, wind farms, and other technologies would be in place. Electricity storage technologies, V2G, and the construction of regional distributed electricity grids would be well underway. Aircraft would begin using biofuels on a significant scale. The transformation of vehicles to using electricity would be well advanced. Plug-in hybrids and all-electric vehicles would be the standard new vehicles being purchased in the latter part of this period.

A decision on whether hydrogen would be a major energy carrier would also be made in this period, after evaluation of the technologies and costs of its production and use based on pilot and large-scale demonstrations. Zero net energy would be achieved for state, local, and federal buildings and by many commercial, residential and industrial buildings and in many communities and areas. Efficiency standards would have been upgraded. It would be routine to make energy-related upgrades to buildings prior to sale.

Other expected features of this period:

- The personal vehicle sector begins a major transformation to electric and plug-in hybrid vehicles as the standard production vehicles.
- Use of IGCC plants running on biomass begins. If not, other modes of deployment of biomass, such as methane production, are put into place.
- Hot rock geothermal energy, wave energy, and other technologies, possibly including carbon sequestration, transition to the commercial stage.

If solar hydrogen or electrolytic hydrogen from wind energy transition to the commercial scale by about 2025, an earlier elimination of CO_2 emissions would be possible. If, on the other hand, some technologies, such as electricity storage from intermediate-scale solar PV, compressed air storage, and V2G do not become commercial, the transition could be delayed. It is not necessary for all these technologies to be commercial, but a combination that would provide for electricity grid reliability on renewable energy alone should exist and be commercial by about 2030. The term "commercial" in this context includes the price that large users of fossil fuels must pay for scarcer CO_2 emission allowances.

Table 8-1 shows the technologies for supply, storage, and conversion, their current status, and the dates when they might come into use in a renewable energy economy, up to about 2025. Table 8-2 shows the same for demand-side technologies.

Table 8-1: Roadmap – Supply and Storage Technologies

Technology	Status	Deployable for large-scale use	Next steps	CO$_2$ abatement cost; obstacles; comments
Solar PV intermediate-scale	Near commercial with time-of-use pricing	2010 to 2015	Orders from industry and government; time-of-use electricity pricing	$10 to $30 per metric ton; no storage; lack of large-scale PV manufacturing (~1 GW/yr/plant); some manufacturing technology development needed.
Solar PV – large-scale	Near commercial	2015 to 2020	Large-scale demonstration with transmission infrastructure, ~5,000 MW by 2015-2020	$20 to $50 per metric ton; no storage; transmission infrastructure may be needed in some cases
Concentrating solar thermal power plants	Near commercial; storage demonstration needed	2015 to 2020	~3,000 to 5,000 MW needed to stimulate demand and demonstrate 12 hour storage, by 2020	$20 to $30 per metric ton in the Southwest. Lack of demand main problem.
Microalgae CO$_2$ capture and liquid fuel production	Technology developed, pilot-scale plants being built	2015	Large-scale demonstrations – 1,000 to 2,000 MW by 2012; nighttime CO$_2$ storage and daytime CO$_2$ capture pilot plants by 2012. Large-scale implementation thereafter. Demonstration plants for liquid fuel production: 2008-2015	Zero to negative at oil prices above $30 per metric ton or so for daytime capture; nighttime capture remains to be characterized. Liquid fuel potential: 5,000 to 10,000 gallons per acre (compared to 650 for palm oil).
Wind power – Large-scale, land-based	Commercial	Already being used	Transmission infrastructure and rules need to be addressed; optimize operation with existing natural gas combined cycle and hydropower plants	Negative to $46 per metric ton for operation with combined cycle standby. Areas of high wind are not near populations. Transmission development needed
Solar PV intermediate storage	Advanced batteries and ultracapacitors are still high cost	~2020	Demonstration of vehicle-to-grid using stationary storage (ultracapacitors and advanced batteries) – several ~1 MW-scale parking lot installations	Five fold cost reduction in stationary storage and lithium-ion batteries needed. Main problems: lack of large-scale manufacturing and some manufacturing technology development needed

Table 8-1 (continued): Roadmap – Supply and Storage Technologies

Technology	Status	Deployable for large-scale use	Next Steps	CO$_2$ abatement cost obstacles; comments
Solar PV intermedi-ate-scale with Vehicle-to-Grid	Planning stage only. Technol-ogy components available. Inte-gration needed.	~2020 to 2025	By 2015, several 5,000 to 10,000 ve-hicle demonstrations of V2G technology	V2G could reduce the cost of solar PV electricity stor-age from several cents to possibly ~1 cent per kWh
Biomass IGCC	Early demonstra-tion stage	~2020	Pilot- and intermedi-ate-scale plants (few MW to 100 MW) with various kinds of biomass (microalgae, aquatic plants), 2015 to 2020	Baseload power
High solar energy capture aquatic biomass	Experience largely in the context of waste-water treatment; some laboratory and pilot plant data	~2020	2010 to 2015 pilot plant evaluations for liquid fuel and meth-ane production with and without connec-tion to wastewater treatment	May be comparable to microalgae biofuels pro-duction. 50 to 100 metric tons per acre
Hot rock geothermal energy	Concept demon-strated; technol-ogy development remains	2025?	Build pilot and demonstration plants: 2015-2020 period	Baseload power
Wave energy	Concepts demon-strated	2020 or 2025?	Pilot and demonstra-tion plants needed	Possible baseload power
Photolytic hydrogen	Laboratory development	Unknown – possibly 2020 or 2025	Significantly in-creased R&D funding, with goal of 2015 pilot plants	Potential for high solar energy capture. Could be a key to overcoming high land-area requirements of most biofuels
Photoelec-trochemical hydrogen	Concept demon-strated; technol-ogy development remains	Possibly 2020 or 2025	Significantly in-creased R&D funding, with goal of 2015 pilot plants	High solar energy capture. Could be a key to overcom-ing problems posed by agri-cultural biofuels (including crop residues)

Table 8-1 (continued): Roadmap – Supply and Storage Technologies

Technology	Status	Deployable for large-scale use	Next Steps	CO_2 abatement cost obstacles; comments
Advanced batteries	Nanotechnology lithium-ion batteries; early commercial stage with subsidies	2015	Independent safety certification (2007?); large-scale manufacturing plants	Large-scale manufacturing to reduce costs. Could be the key to low cost V2G technology
Carbon sequestration	Technology demonstrated in context other than power plants	Unknown. Possibly 15 to 20 years.	Long-term leakage tests. Demonstration project ~2015-2020	For use with biomass, plus back up, if coal is needed
Ultracapacitors	Commercial in certain applications but not for large-scale energy storage	2015 to 2020?	Demonstration test with intermediate-scale solar PV. Demonstrate with plug-in hybrid as a complement to battery operation for stop-and-start power	Complements and tests V2G technology. Significant cost reduction needed for cost to be ~$50/metric ton CO_2. Lower CO_2 price with time-of-use rates
Nanocapacitors	Laboratory testing of the concepts	Unknown.	Complete laboratory work and demonstrate the approach	Has the potential to reduce costs of stationary electricity storage and take ultracapacitor technology to the next step
Electrolytic hydrogen production	Technology demonstrated	Depends on efficiency improvements and infrastructure development	Demonstration plant with compressed hydrogen vehicles needed ~2015-2020	Could be used in conjunction with off-peak wind power

Table 8-2: Roadmap – Demand-Side Technologies, 2008-2020

Technology	Status	Deployable for large-scale use	Next steps	CO$_2$ price; obstacles; comments
Efficient gasoline and diesel passenger vehicles	Commercial to ~40 miles per gallon or more	Being used	Efficiency standards needed	Efficiency depends on the vehicle. Can be much higher.
Plug-in hybrid vehicles	Technology has been demonstrated	2010	Efficiency standards, government and corporate orders for vehicles	Large-scale battery manufacturing needed to reduce lithium-ion battery cost by about a factor of five.
Electric cars	Technology with ~200 mile range has been demonstrated; low volume commercial production in 2007 (sports car and pickup truck)	2015 to 2020	Safety testing, recycling infrastructure for battery materials, large-scale orders, solar PV-V2G demonstration	One of the keys to reducing the need for biofuels and increasing solar and wind power components.
Internal combustion hydrogen vehicles	Technology demonstrated	Depends on infrastructure development	10,000 psi cylinder development and testing of vehicles. Demonstration project	
Biofuels for aircraft	Various fuels being tested	2020?	Fuel development, safety testing, emissions testing	
Hydrogen-fuel aircraft	Technology has been demonstrated	2030?	Aircraft design, safety testing, infrastructure demonstration	In combination with solar hydrogen production, could reduce need for liquid biofuels.
Building design	Commercial, well known	Already being used	Building standards, dissemination of knowledge, elimination of economic disconnect between building developers and users	Residential and commercial building energy use per square foot can be reduced 60 to 80 percent with existing technology and known approaches. CO$_2$ price, negative to $50 per metric ton.
Geothermal heat pumps	Commercial	Already being used	Building standards that specify performance will increase its use	Suitable in many areas; mainly for new construction.

Table 8-2 (continued): Roadmap – Demand-Side Technologies, 2008-2020

Technology	Status	Deployable for large-scale use	Next steps	CO$_2$ price; obstacles; comments
Combined heat and power (CHP), commercial buildings and industry	Commercial	Already being used	Building performance standards and CO$_2$ cap will increase use	CO$_2$ price negative to <$30 per metric ton in many circumstances.
Micro-CHP	Semi-commercial	Already being used	Building performance standards will increase use	
Compact fluorescent lighting (CFL)	Commercial	Being used currently	Appliance and building regulations needed	Negative CO$_2$ price. Mercury impact of disposal needs to be addressed.
Hybrid solar light-pipe and CFL	Technology demonstrated; beta-testing being done in commercial establishments	2012 to 2015?	Government and commercial sector orders	Solar concentrators focus light indoors; work in conjunction with CFL. Five-fold cost reduction needed.
Industrial sector: examples of technologies and management approaches: alternatives to distillation, steam system management, CHP, new materials, improved proportion of first pass production	Constant development of processes	Various	Hard cap for CO$_2$ with annual assured decreases and no free allowances will lead to increase in efficiency	Variable. Negative to possibly $50 per metric ton, possibly more in some cases. Great potential for economical increases in efficiency exists at present costs, since energy costs have gone up suddenly. Successful reductions of energy use indicate that overall cost will be modest, with possible reduction in net cost of energy services.

C. Macroeconomics of the Transition

In the three decades following 1970, U.S. energy expenditures fluctuated from a low of about six percent (very briefly when prices collapsed in the late 1990s) to about 14 percent of the GDP. About 8 percent has been more typical, leaving aside the fluctuations caused by the turbulence immediately following the crises of 1973 and 1979. The proportion fell briefly to about 6 percent in the late 1990s, when oil prices declined steeply, dipping to a low of $12 per barrel.

Figure 8-1: Proportion of GDP Spent on Energy

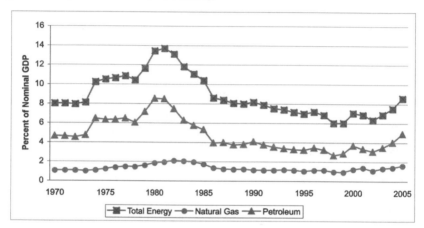

Source: Courtesy of the Energy Information Administration of the United States Department of Energy

By 2050 the GDP will be nearly $40 trillion (constant 2004 dollars) under business-as-usual economic growth.[4] The energy use projected under the business-as-usual scenario is 160 quadrillion Btu, while that estimated for the reference scenario for the present analysis is about 76 quadrillion Btu. Both figures include losses in electricity production; the latter also includes losses in biofuels production. (The energy consumption in 2005 was about 100 quadrillion Btu.)

We have estimated the proportion of GDP that would be devoted to the energy services, such as transportation and heating and cooling in buildings. One overall criterion for an economical transition to a renewable energy economy is that the proportion of GDP devoted to energy services be no different than has been typical in recent decades, apart from the brief extreme swings occasioned by very rapid increases and decreases of oil prices. It is more difficult to compare this macroeconomic estimate for the reference scenario with the proportion of GDP that would be devoted to energy under the business-as-usual scenario. For the purposes of comparison, we use present prices, though this represents a rather unrealistic picture. The reason is that such a projection is built into a business-as-usual scenario, which is less a projection than an estimate of energy use in the future in the absence of major changes in the global economic, political, security, and resource picture. We chose a benchmark eight percent of GDP for energy expenses as a figure of merit for the reference scenario. A comparison with business-as-usual is made under assumption of present energy prices.[5] We address issues connected with business-as-usual projections separately (see Section C below).

1. The Residential and Commercial Sectors

A computation of the future cost of energy services under the reference scenario requires estimates of energy supply costs (fuel and electricity) and of additional investments that will be necessary to achieve the higher efficiency relative to the business-as-usual scenario.

Present costs of ethanol, hydrogen from electrolysis, and other biofuels indicate that the costs of biofuel supply for the residential and commercial sectors may be somewhat higher in the future than that of fossil fuels in 2005. We have assumed a delivered cost of $20 per million Btu, which is rather on the pessimistic side, in order not to underestimate the future fuel cost in a reusable energy economy.

For electricity, we assume a delivered cost to residential and commercial customers of about 12 cents per kWh for two-thirds of the supply, based on IGCC technology with sequestration and coal as a fuel, with which much of the future renewable electric supply system would have to compete in the absence of subsidies. For the rest, we have assumed that the cost would be typical of an intermediate-level solar PV system. We also assume that storage corresponding to one day's average output would be part of such a system. Storage capacity costs are taken to be $200 per kWh, which is about one-fifth the present price of ultra-capacitors.[6] The installed cost of solar PV systems is assumed to average $1.50 per peak watt, without storage. The generation per peak installed kW is taken as 1,800 kWh per year for a non-tracking system. A two-cent charge for distribution is added, since distribution systems will likely have to be strengthened for widespread use of intermediate-scale solar PV systems. The overall cost for such a system comes to about 18.2 cents per kWh. Combining the two estimates yields an average electricity cost for the residential and commercial sectors of 14.1 cents per kWh. Other forms of storage could be used instead or as complements in a "smart grid" system that combines supply-side and demand-side storage.[7]

For the business-as-usual scenario, we have used January 2006 costs: $12 and $10 per million Btu for the residential and commercial sectors respectively for fuel, and 9.57 cents and 8.81 cents per kWh for electricity. As discussed above, these are only notional costs used here to represent an unchanged and smooth business-as-usual energy future.[8] They are unlikely to be representative of actual future costs if energy demand grows as estimated in the business-as-usual scenario. Increasing fuel consumption implies growing imports of oil and natural gas (See Section C below), which will likely affect market and geopolitical conditions adversely.

We also assume that additional investments will be needed relative to business-as-usual to achieve the efficiencies that are built into the demand structure in the reference scenario. It is more difficult to make reliable estimates of such investments far into the future in part because there are fewer generally applicable examples.

1. For new commercial buildings, the added investment assumed is $10 per square foot, which is greater than examples of platinum level LEED-certified buildings. LEED (Leadership in Energy and Environmental Design) is a building certification program that evaluates not only energy efficiency but also other environmental aspects such as water use and the nature of the materials used on construction. We have not attributed any of the costs to aspects of environmental design other than energy use.

2. Residential building costs are much more variable, varying from $70 to over $200 per square foot for environmentally advanced buildings. There is no discernible pattern, except that buildings that include solar PV, solar thermal space or water heating, or geothermal heat pumps would cost somewhat more. (see Table 8-3). We assumed that the higher efficiency in the reference scenario would add about 10 percent per square foot to the cost of advanced buildings being built at present, as illustrated in Table 8-3. Only costs for efficiency improvements are included. The costs for solar PV, solar thermal installations, and combined heat and power systems were added separately.

3. For existing buildings, we assumed an investment at the time of sale of the homes and a turn over rate of a little over 5 percent per year. The total sales of existing homes between 2010 and 2050 would be about 300 million (since existing homes would be sold more than once in the period). We assumed that there would be an investment of $20,000 in one-third of these transactions.

Table 8-3: Examples of Cost of Green Building Award-Winning Homes for Efficiency Improvements Only

Climate/State	cost/sq. ft.	area, sq. ft.	Cost $
Moderate/MD or VA	100	1900	190000
Cold/WI	76	2728	207328
Hot/TX	115	1994	224310
Moderate/CA	70	2543	163610
Cold/CO	98	2864	280672
Cold/MI	198	3453	676194
Cold/ID	75	2653	198975
Moderate/MD	58	3716	192128
Moderate/OR	235	2544	565540
Total		24395	2698757
Average	111		

Source: Energy Value Housing Awards at http://www.nahbrc.org/evha/winners.html (EVHA 2007) and, for the first building in the list at PRSEA 2003.
Note: The additional costs of solar thermal installations over and above those of conventional systems are taken to be: solar PV at $6,000 per peak watt, solar thermal water heating systems at $5,000, and geothermal heat pumps at $7,500 for those homes that have them. These costs have been subtracted from the building cost and separately accounted for in the reference scenario and Table 8-4 below.

Table 8-4 shows the results for the residential and commercial sectors. The total estimated annual energy and investment costs for the residential and commercial sectors in terms of GDP impact are about the same as energy costs in the business-as-usual scenario. The lower per house and per square foot, higher needed investment, and higher anticipated per unit costs of electricity and fuels under the IEER reference scenario are taken into account. The net estimated GDP impact of reducing residential and commercial sector energy use by efficiency improvements and converting entirely to renewable energy sources is small and well within the range of the uncertainties in the calculations.

Table 8-4: Annual Residential (R) and Commercial (C) Energy and Investment Costs in 2050, in Billions of Constant 2005 Dollars

Item	IEER Reference Scenario	Business-as-Usual Scenario
R + C Electricity	$326	$442
R + C Fuel	$150	$247
Sub-total energy cost	$476	$689
Added annual investment for efficiency (Notes 2 and 3)	$205	$0
Total GDP-basis amount (rounded)	*$681*	*$689*
GDP in 2050 (Note 4)	$40,000	$40,000
GDP fraction: residential and commercial energy services	*1.70%*	*1.72%*

Notes:
1. Business-as-usual (BAU) fuel and electricity prices: about $12 per million Btu and 9.6 cents per kWh. Reference Scenario prices: $20 per million Btu and 14.1 cents per kWh respectively. BAU electricity price is from January 2006.
2. Added efficiency investments: existing residences: $20,000 per residence each time, assumed to occur in one of every three sales of existing buildings between 2010 and 2050; new = $10 per square foot (about $20,000 per house, approximate LEED-certified house added cost); plus cost of replacing appliances every 15 years with then-prevailing advanced appliances. Investments for solar thermal heating, combined heat and power, and geothermal heat pumps added to these figures for the proportion of residential area using them. LEED stands for Leadership in Energy and Environmental Design; it is a building certification program.
3. Commercial efficiency investments: $10 per square foot; this is more than examples of platinum level LEED investment. Investments for solar thermal heating, combined heat and power, and geothermal heat pumps have been added to these figures.
4. GDP = consumption expenditures + investment + government spending (on goods and services) + exports – imports.

Under the stated assumptions, the costs in the residential sector are somewhat higher than business-as-usual and those in the commercial sector are somewhat lower. A calculation for an average individual homeowner who purchases a new, detached home in the year 2050, with features weighted by the proportion in which they are used in the reference scenario indicates that the added cost would be $20 to $100 per month. An interest rate of 7 percent and a 30-year mortgage has been assumed. The latter figure is less than 0.7 percent of median household income in 2050. The range reflects uncertainties as to the marginal increased cost of efficiency based on estimated added costs of efficient homes over typical homes at present of 3 to 8 percent.[9]

2. Transportation

Estimating the costs of the transformation of the vehicular sector for the technologies in the reference scenario is rather difficult and relies on a projection of the costs of plug-in hybrids and electric vehicles. The most important uncertainty is the cost of batteries. At present the cost is around $1,000 per kWh. This is too expensive to compete with gasoline cars at $3 per gallon. However, as noted, present battery costs are dominated by low volume of manufacture and the nascent nature of the industry. We assume battery costs of $200 per kWh, which are anticipated in less than a decade (see Chapters 3 and 5). We also assume that the entire cost of the battery needed for a 200-mile range would be additional cost over a gasoline car. Efficiency assumptions for the year 2050 for personal vehicles are as follows:

- Business-as-usual: about 40 miles per gallon.
- IEER reference scenario: 10 miles per kWh
- An average electricity cost of 14.1 cents per kWh, assuming that partial off-peak and partial on-peak charging will result in average electricity rates for vehicle charging. This assumption may appear rather adverse for electric cars. However, it is realistic to assume that facilities similar to gas stations would be commonly used for quick charging of vehicles in addition to off-peak charging in a context where electric vehicles and/ or plug-in hybrids with high capacity for running on electricity only would be the standard vehicles on the market.

The reduced costs of maintenance (no oil changes, no tune-ups, lower brake replacement rate, etc.) of electric vehicles are not taken into account. With these assumptions, the proportion of GDP devoted to fuel cost for personal vehicles would be about 0.9 percent for the business-as-usual scenario and 0.5 to 0.6 percent for the reference scenario. Another way to look at these numbers is that personal and small business transportation in the reference scenario would be comparable to the business-as-usual scenario with present achievable electric vehicle efficiency and battery cost of $200 per kWh. At future efficiency of 10 miles per kWh, the battery cost could be about $400 per kWh. Hence, improvements in vehicle efficiency and reductions in battery costs can go hand-in-hand in improving electric vehicle economics.

Personal transportation fuel use represents only about half the fuel consumption in transportation. The proportion of energy costs in the transportation sector would therefore be 2 to 3 percent, possibly less, under these assumptions in the year 2050.

D. Projecting Business-as-usual

A business-as-usual future would be characterized by a lack of restrictions on fossil fuel consumption and hence most likely growing oil and natural gas imports. Such an energy future may be characterized by economic turbulence and higher prices that are not captured by the notional prices used in the comparisons above. Business-as-usual is an historical construct that facilitates technical calculations, but should not be regarded as an estimate of the evolution of the energy future of the United States or the world.

An energy future that follows the past pattern of increasing oil imports would likely be wracked by volatility in oil prices. Disruptions in supply, such as those caused by Hurricane Katrina, may also be more frequent due to the increasing effects of severe climate change. If the United States does not commit to serious reductions in oil consumption, there would be no prospect that China, India, and other developing countries would do so. The overall global economic and political environment in which these and other countries, including the European Union and Japan, compete for oil and gas would be very likely to deteriorate. This problem of resource competition would likely be much worse in areas where production costs are very low, at present mainly the Persian Gulf region, where costs are less than $3 per barrel, but also in other areas, where production costs are moderate.

Another way of saying the same thing is that business-as-usual projections of energy use are unlikely, in the same way that projections made before 1973 became unlikely in the face of the political, military, and economic crisis represented by the events of 1973 and 1979. They changed the energy picture in the United States profoundly (see Chapter 1). The main choice is whether energy use will become more efficient and more oriented towards domestic renewable resources by deliberate policy or whether it will be driven there willy-nilly by recurrent global crises.

CHAPTER 9: SUMMARY

A three-fold global energy crisis has emerged since the 1970s; it is now acute on all three fronts:

1. **Climate disruption**: Carbon dioxide (CO_2) emissions due to fossil fuel combustion are the main anthropogenic cause of severe climate disruption, whose continuation portends grievous, irreparable harm to the global economy, society, and current ecosystems.
2. **Insecurity of oil supply**: Rapid increases in global oil consumption and conflict in and about oil exporting regions make prices volatile and supplies insecure.
3. **Nuclear proliferation:** Non-proliferation of nuclear weapons is being undermined in part by the spread of commercial nuclear power technology, which is being put forth as a major solution for reducing CO_2 emissions.

This book examines the technical and economic feasibility of achieving a U.S. economy with zero-CO_2 emissions without nuclear power. This is interpreted as an elimination of all but a few percent of CO_2 emissions or complete elimination with the possibility of removing from the atmosphere some CO_2 that has already been emitted. We set out to answer three questions:

- Is it possible to physically eliminate CO_2 emissions from the U.S. energy sector without resort to nuclear power, which has serious security and other vulnerabilities?
- Is a zero-CO_2 economy possible without purchasing offsets from other countries – that is, without purchasing from other countries the right to continue emitting CO_2 in the United States?
- Is it possible to accomplish the above at reasonable cost?

The overarching finding of this study is that a zero-CO_2 U.S. economy can be achieved within the next thirty to fifty years without the use of nuclear power and without acquiring carbon credits from other countries. In other words, actual physical emissions of CO_2 from the energy sector can be eliminated with technologies that are now available or foreseeable. This can be done at reasonable cost while creating a much more secure energy supply than at present. Net U.S. oil imports can be eliminated in about 25 years. All three insecurities – severe climate disruption, oil supply and price insecurity, and nuclear proliferation via commercial nuclear energy – will thereby be addressed. In addition, there will be large ancillary health benefits from the elimination of most regional and local air pollution, such as high ozone and particulate levels in cities, which is due to fossil fuel combustion.

The achievement of a zero-CO_2 economy without nuclear power will require unprecedented foresight and coordination in policies from the local to the national, across all sectors of the energy system. Much of the ferment at the state and local level, as well as some of the proposals in Congress, is already pointed in the right direction. But a clear long-term goal is necessary to provide overall policy coherence and establish a yardstick against which progress can be measured.

A zero-CO_2 U.S. economy without nuclear power is not only achievable – it is necessary for environmental protection and security. *Even the process of the United States setting a goal of a zero-CO_2, nuclear-free economy and taking initial firm steps towards it will transform global energy politics in the immediate future and establish the United States as a country that leads by example, rather than one that preaches temperance from a barstool, especially in the matter of nuclear power and the technologies that are associated with it, some of which are directly relevant to nuclear weapons production.*

A. Findings

Finding 1: A goal of a zero-CO_2 economy is necessary to minimize harm related to climate change.

According to the Intergovernmental Panel on Climate Change, global CO_2 emissions would need to be reduced by 50 to 85 percent relative to the year 2000 in order to limit average global temperature increase to 2 to 2.4 degrees Celsius relative to pre-industrial times. A reduction of 80 percent in total U.S. CO_2 emissions by 2050 would be entirely inadequate to meet this goal. It implies annual U.S. emissions of about 2.8 metric tons per person.

A global norm of emissions at this rate would leave worldwide CO_2 emissions almost as high as in the year 2000.[1] In contrast, if a global norm of approximately equal per person emissions by 2050 is created along with a 50 percent global reduction in emissions, it would require an approximately 88 percent reduction in U.S. emissions. An 85 percent global reduction in CO_2 emissions corresponds to a 96 percent reduction for the United States. An allocation of emissions by the standard of cumulative historical contributions would be even more stringent.

A U.S. goal of zero-CO_2, defined as being a few percent on either side of zero relative to 2000, is both necessary and prudent for the protection of global climate. It is also implied by the United Nations Framework Convention on Climate Change. That treaty, to which the United States is a party, requires that the burden of reducing emissions of greenhouse gases be shared equitably, with due consideration to the historical fact and current reality that developed countries have been and are responsible for most emissions. A per-capita norm is a minimal interpretation of this treaty. When joined to the goal of being reasonably sure to limit temperature rise to the range of 2 to 2.4 degrees Celsius by 2050, the UNFCCC implies a zero-CO_2 economy for the United States.

Finding 2: A hard cap on CO_2 emissions – that is, a fixed emissions limit that declines year by year until it reaches zero – would provide large users of fossil fuels with a flexible way to phase out CO_2 emissions. However, free allowances, offsets that permit emissions by third party reductions,[2] or international trading of allowances, notably with developing countries that have no CO_2 cap, would undermine and defeat the purpose of the system. A measurement-based physical limit, with appropriate enforcement, should be put into place.

A hard cap on CO_2 emissions is recommended for large users of fossil fuels, defined as an annual use of 100 billion British thermal units (Btu) or more – equal to the delivered energy use of about 1,000 households. At this level, users have the financial resources to be able to track the market, make purchases and sales, and evaluate when it is most beneficial to invest in CO_2 reduction technologies relative to purchasing credits. This would cover about two-thirds of fossil fuel use. Private vehicles, residential and small commercial use of natural gas and oil for heating, and other similar small-scale uses would not be covered by the cap. The transition in these areas would be achieved through efficiency standards, tailpipe emissions standards, and other standards set and enforced by federal, state, and local governments. Taxes are not envisaged in this study, except possibly on new vehicles that fall far below the average efficiency or emissions standards. The hard cap would decline annually and be set to go to zero before 2060. Acceleration of the schedule would be possible, based on developments in climate impacts and technology.

The annual revenues that would be generated by the government from the sale of allowances would be on the order of $30 billion to $50 billion per year through most of the period, since the price of CO_2 emission allowances would tend to increase as supply goes down. These revenues would be devoted to ease the transition at all levels – local, state, and federal – as well as for demonstration projects and research and development.

Finding 3: A reliable U.S. electricity sector with zero-CO_2 emissions can be achieved without the use of nuclear power or fossil fuels.

The U.S. renewable energy resource base is vast and practically untapped. Available wind energy resources in 12 Midwestern and Rocky Mountain states equal about 2.5 times the entire electricity production of the United States. North Dakota, Texas, Kansas, South Dakota, Montana, and Nebraska each have wind energy potential greater than the electricity produced by all 103 U.S. nuclear power plants. Solar energy resources on just one percent of the area of the United States are about three times as large as wind energy, if production is focused in the high insolation areas in the Southwest and West.

Just the parking lots and rooftops in the United States could provide most of the U.S. electricity supply. This also has the advantage of avoiding the need for transmission line expansion, though some strengthening of the distribution infra-structure may be needed. Wind energy is already more economical than nuclear power. In the past two years, the costs of solar cells have come down to the point that medium-scale installations, such as the ones shown in Chapter 3, are economical in sunny areas, since they supply electricity mainly during peak hours.

The main problem with wind and solar energy is intermittency. This can be re-duced by integrating wind and solar energy together into the grid – for instance, wind energy is often more plentiful at night. Geographic diversity also reduces the intermittency of each source and for both combined. Integration into the grid of these two sources up to about 15 percent of total generation (not far short of the contribution of nuclear electricity today) can be done without serious cost or technical difficulty with available technology, provided appropriate optimization steps are taken.

Solar and wind should also be combined with hydropower – with the latter being used when the wind generation is low or zero. This is already being done in the Northwest. Conflicts with water releases for fish management can be addressed by combining these three sources with natural gas standby. The high cost of natural gas makes it economical to use combined cycle power plants as standby capacity and spinning reserve for wind rather than for intermediate or baseload generation. In other words, given the high price of natural gas, these plants could be economically idled for some of the time and be available as a complement to wind power. Compressed air can also be used for energy storage in combination with these sources. No new technologies are required for any of these generation or storage methods.

Baseload power can be provided by geothermal and biomass-fueled generat-ing stations. Intermediate loads in the evening can be powered by solar thermal power plants which have a few hours of thermal energy storage built in.

Finally, new batteries can enable plug-in hybrids and electric vehicles owned by fleets or parked in large parking lots to provide relatively cheap storage. Nano-technology-based lithium-ion batteries, which Altairnano has begun to produce, can be deep discharged far more times than needed simply to operate the vehicle over its lifetime (10,000 to 15,000 times compared to about 2,000 times respectively).

Since the performance of the battery is far in excess of the cycles of charging and discharging needed for the vehicle itself, vehicular batteries could become a very low-cost source of electricity storage that can be used in a vehicle-to-grid (V2G) system. In such a system, parked cars would be connected to the grid and charged and discharged according to the state of the requirements of the grid and the charge of the battery in the vehicle. Communications technology to accomplish this via wires or wireless means is already commercial. A small fraction of the total number of road vehicles (several percent) could provide sufficient backup capacity to stabilize a well designed electricity grid based on renewable energy sources (including biomass and geothermal).

One possible configuration of the electric power grid is shown in Figure 5-6 in Chapter 5. A large amount of standby power is made available. This allows a combination of wind and solar electricity to supply half or more of the electricity without affecting reliability. Most of the standby power would be supplied by stationary storage and/or V2G and by combined cycle power plants for which the fuel is derived from biomass. Additional storage would be provided by thermal storage associated with central station solar thermal plants. Hydropower use would be optimized with the other sources of storage and standby capacity. Wind energy can also be complemented by compressed air storage, with the compressed air being used to reduce methane consumption in combined cycle power plants. Storage on the energy supply-side can be combined with storage on the demand-side and a smart grid approach in which demand can be adjusted to more closely match renewable energy supply.

With the right combination of technologies, it is likely that even the use of coal can be phased out, along with nuclear electricity. However, we recognize that the particular technologies that are on the cutting edge today may not develop as now appears likely. It therefore appears prudent to have a backup strategy. The carbon dioxide from coal-fired power plants can be captured at moderate cost if the plants are used with a technology called integrated gasification combined cycle (IGCC). Carbon capture and sequestration may also be needed for removing CO_2 from the atmosphere via biomass.

Finding 4: The use of nuclear power entails risks of nuclear proliferation, terrorism, and serious accidents. It exacerbates the problem of nuclear waste and perpetuates vulnerabilities and insecurities in the energy system that are avoidable.

Commercial nuclear technology is being promoted as a way to reduce CO_2 emissions, including by the U.S. government. With Russia, the United States has also been promoting a scheme to restrict commercial uranium enrichment and plutonium separation (reprocessing) to the countries that already have it. (These are both processes that can produce nuclear-weapons-usable materials.) This is a transparent attempt to change the Nuclear Non-Proliferation Treaty (NPT) without going through the process of working with the signatories to amend it. The effort will undermine the treaty, which gives non-nuclear parties an "inalienable right" to commercial nuclear technology. In any case, non-nuclear-weapon states are unlikely to go along with the proposed restrictions.

It is not hard to discern that the increasing interest in nuclear power is at least partly as a route to acquiring nuclear weapons capability. For instance, the Gulf Cooperation Council (Bahrain, Kuwait, Oman, Qatar, Saudi Arabia and the United Arab Emirates), pointing to Iran and Israel, has stated that it will openly acquire civilian nuclear power technology. In making the announcement, the Saudi Foreign Minister Prince Saud Al-Faisal was quoted in the press as saying "It is not a threat….We are doing it openly." He also pointed to Israel's nuclear reactor, used for making plutonium for its nuclear arsenal, as the "original sin." At the same time, he urged that the region be free of nuclear weapons.[3]

Interest in commercial reprocessing may grow as a result of U.S. government policies. The problems of reprocessing are already daunting. For instance, North Korea used a commercial sector power plant and a reprocessing plant to get the plutonium for its nuclear arsenal. Besides the nuclear weapon states, about three dozen countries, including Iran, Japan, Brazil, Argentina, Egypt, Taiwan, South Korea, and Turkey, have the technological capacity to make nuclear weapons. It is critical for the United States to lead by example and achieve the necessary reductions in CO_2 emissions without resorting to nuclear power. Greater use of nuclear power would convert the problem of nuclear proliferation from one that is difficult today to one that is practically intractable.

Even the present number of nuclear power plants and infrastructure has created tensions between non-proliferation and the rights countries have under the NPT to acquire commercial nuclear technology. Increasing their number would require more uranium enrichment plants, when just one such plant in Iran has stoked global political-security tensions to a point that it is a major driver in spot market oil price fluctuations. In addition, there are terrorism risks, since power plants are announced terrorist targets. It hardly appears advisable to increase the number of targets.

The nuclear waste problem has resisted solution. Increasing the number of power plants would only compound the problem. In the United States, it would likely create the need for a second repository, and possibly a third, even though the first, at Yucca Mountain in Nevada, is in deep trouble. No country has so far been able to address the significant long-term health, environmental and safety problems associated with spent fuel or high level waste disposal, even as official assessments of the risk of harm from exposure to radiation continue to increase.

Finally, since the early 1980s, Wall Street has been, and remains, skeptical of nuclear power due to its expense and risk. That is why, more than half a century after then-Chairman of the Atomic Energy Commission, Lewis Strauss, proclaimed that nuclear power would be "too cheap to meter," the industry is still turning to the government for loan guarantees and other subsidies. The insurance side is no better. The very limited insurance that does exist is far short of official estimates of damage that would result from the most serious accidents; it is almost all government-provided.

Finding 5: The use of highly efficient energy technologies and building design, generally available today, can greatly ease the transition to a zero-CO_2 economy and reduce its cost. A two percent annual increase in efficiency per unit of Gross Domestic Product relative to recent trends would result in a one percent decline in energy use per year, while providing three percent GDP annual growth. This is well within the capacity of available technological performance.

Before the first energy crisis in 1973, it was generally accepted that growth in energy use and economic growth, as expressed by Gross Domestic Product (GDP), went hand in hand. But soon after, the U.S. energy picture changed radically and economic growth was achieved for a decade without energy growth.

Since the mid-1990s, the rate of energy growth has been about two percent less than the rate of GDP growth, despite the lack of national policies to greatly increase energy efficiency. For instance, residential and commercial buildings can be built with just one-third to one-tenth of the present-day average energy use per square foot with existing technology. As another example, we note that industrial energy use in the United States has stayed about the same since the mid-1970s, even as production has increased.

Our research indicates that annual use of delivered energy (that is, excluding energy losses in electricity and biofuels production) can be reduced by about one percent per year while maintaining the economic growth assumed in official energy projections.

Finding 6: Biofuels, broadly defined, could be crucial to the transition to a zero-CO_2 economy without serious environmental side effects or, alternatively, they could produce considerable collateral damage or even be very harmful to the environment and increase greenhouse gas emissions. The outcome will depend essentially on policy choices, incentives, and research and development, both public and private.

Food crop-based biodiesel and ethanol can create and are creating social, economic, and environmental harm, including high food prices, pressure on land used by the poor in developing countries for subsistence farming or grazing, and emissions of greenhouse gases that largely or completely negate the effect of using the solar energy embodied in the biofuels. While they can reduce imports of petroleum, ethanol from corn and biodiesel from palm oil are two prominent examples of damaging biofuel approaches that have already created such problems even at moderate levels of production.

For instance, in the name of renewable energy, the use of palm oil production for European biodiesel use has worsened the problem of CO_2 emissions due to fires in peat bogs that are being destroyed in Indonesia, where much of the palm oil is produced. Rapid increases in ethanol from corn are already partly responsible for fueling increases in tortilla prices in Mexico. Further, while ethanol from corn would reduce petroleum imports, its impact on reducing greenhouse gas emissions would be small at best due to the energy intensity of both corn and ethanol production, as well as the use of large amounts of artificial fertilizers, which also result in emissions of other greenhouse gases (notably nitrous oxide). All subsidies for fuels derived from food crops should be eliminated.

In contrast, biomass that has high efficiency solar energy capture (~five percent), such as microalgae grown in a high-CO_2 environment, can form a large part of the energy supply both for electricity production and for providing liquid and gaseous fuels for transport and industry. Microalgae have been demonstrated to capture over 80 percent of the daytime CO_2 emissions from power plants and can be used to produce up to 10,000 gallons of liquid fuel per acre per year. Some aquatic plants, such as water hyacinths, have similar efficiency of solar energy capture and can be grown in wastewater as part of combined water treatment and energy production systems.

Water hyacinths have been used to clean up wastewater because they grow rapidly and absorb large amounts of nutrients. Their productivity in tropical and subtropical climates is comparable to microalgae – up to 250 metric tons per hectare per year. They can be used as the biomass feedstock for producing liquid and gaseous fuels. There are also other high productivity aquatic plants, such as duckweed, that grow in a wider range of climates that can be used for producing biofuels.

Prairie grasses have medium productivity, but can be grown on marginal lands in ways that allow carbon storage in the soil. This approach can therefore be used both to produce fuel renewably and to remove CO_2 from the atmosphere.

Finally, solar energy can be used to produce hydrogen; this could be very promising for a transition to hydrogen as a major energy source. Techniques include photoelectrochemical hydrogen production using devices much like solar cells, high-temperature, solar-energy-driven splitting of water into hydrogen and oxygen, and conversion of biomass into carbon monoxide and hydrogen in a gasification plant.

Finding 7: Much of the reduction in CO_2 emissions can be achieved without incurring any cost penalties (as, for instance, with efficient lighting and refrigerators). The cost of eliminating the rest of CO_2 emissions due to fossil fuel use is likely to be in the range of $10 to $30 per metric ton of CO_2.

Table 9-1 shows the estimated costs of eliminating CO_2 from the electricity sector using various approaches.

Table 9-1: Summary of costs for CO_2 abatement (and implicit price of CO_2 emission allowances) – Electricity sector (based on 2004 costs of energy)

CO_2 source	Abatement method	Phasing	Cost per metric ton CO_2, $	Comments
Pulverized coal	Off-peak wind energy	Short-term	A few dollars to $15	Based on off-peak marginal cost of coal.
Pulverized coal	Capture in micro-algae	Short- and medium-term	Zero to negative	Assuming price of petroleum is >$30 per barrel.
Pulverized coal	Wind power with natural gas standby	Medium- and long-term	Negative to $46	Combined cycle plant idled to provide standby. Highest cost at lowest gas price: $4/mn Btu
Pulverized coal	Nuclear power	Medium- to long-term	$20 to $50	Unlikely to be economical compared to wind with natural gas standby.
Pulverized coal	Integrated Gasification Combined Cycle (IGCC) with sequestration	Long-term	$10 to $40 or more	Many uncertainties in the estimate at present. Technology development remains.
Natural gas standby component of wind	Electric vehicle-to-grid	Long-term	Less than $26	Technology development remains. Estimate uncertain. Long-term-natural gas price: $6.50 per million Btu or more.

Notes:

1. Heat rate for pulverized coal = 10,000 Btu/kWh; for natural gas combined cycle = 7,000 Btu/kWh.

2. Wind-generated electricity costs = 5 cents per kWh; pulverized coal = 4 cents per kWh; nuclear = 6 to 9 cents per kWh.

3. Petroleum costs $30 per barrel or more.

4. CO_2 costs associated with wind energy related items can be reduced by optimized deployment of solar and wind together.

Further, the impact of increases in costs of CO_2 abatement on the total cost of energy services is low enough that the overall share of GDP devoted to such services would remain at about the present level of about 8 percent or perhaps decline. It has varied mainly between 8 and 14 percent since 1970, hitting a peak in 1980. It dropped briefly to about 6 percent in the late 1990s when oil prices tumbled steeply, hitting a low of about $12 per barrel in 1998.

Finding 8: The potential for energy efficiency is considerably greater than assumed in the reference scenario in many areas. Greater efficiency, greater use of electricity, and use of hydrogen derived from wind (and possibly solar) energy would greatly reduce the land impacts associated with large-scale biofuel production.

The opportunities for greater efficiency beyond the reference scenario discussed in Chapter 6 help reduce the requirement for liquid and gaseous biofuels in 2050 from about 35 quadrillion Btu to 20 to 25 quadrillion Btu. A significant fraction of this fuel requirement can be met by electrolytic hydrogen from wind and possibly direct solar hydrogen production, provided there is adequate early emphasis on commercialization of hydrogen. Distributed hydrogen production and use of hydrogen in internal combustion engines are the closest to practical application. Reducing liquid and gaseous biofuels requirements to the 10 to 15 quadrillion Btu range would largely resolve the most important anticipated environmental impact of the reference scenario – land use for biofuels. In the preferred renewable future, only about 2 to 3 percent of the land area of the U.S. would be needed for energy supply.

Finding 9: The transition to a zero-CO_2 system can be made in a manner compatible with local economic development in areas that now produce fossil fuels.

Fossil fuels are mainly produced today in the Appalachian region, in the Southwest and West and some parts of the Midwest and Rocky Mountain states. These areas are also well-endowed with the main renewable energy resources – solar and wind. Federal, state and regional policies, designed to help workers and communities transition to new industries, therefore appear to be possible without more major physical movement or disruption of populations than has occurred in post-World War II United States. It is recognized that much of that movement has been due to dislocation and shutdown of industries, which causes significant hardship to communities and workers. Some of the resources raised by the sale of CO_2 allowances should be devoted to reducing this disruption. For instance, the use of CO_2 capture technologies, notably microalgae CO_2 capture from existing fossil fuel plants, can create new industries and jobs in the very regions where the phase-out of fossil fuels would have the greatest negative economic impact. Public policy and direction of financial resources can help ensure that new energy sector jobs that pay well are created in those communities.

B. Recommendations: The Clean Dozen

The 12 most critical policies that need to be enacted as urgently as possible for achieving a zero-CO_2 economy without nuclear power are as follows.

1. Enact a physical limit of CO_2 emissions for all large users of fossil fuels (a "hard cap") that steadily declines to zero prior to 2060, with the time schedule being assessed periodically for tightening according to climate, technological, and economic developments. The cap should be set at the level of some year prior to 2007, so that early implementers of CO_2 reductions benefit from the setting of the cap. Emission allowances would be sold by the U.S. government for use in the United States only. There would be no free allowances, no offsets and no international sale or purchase of CO_2 allowances. The estimated revenues – approximately $30 to $50 billion per year – would be used for demonstration plants, research and development, and worker and community transition.
2. Eliminate all subsidies and tax breaks for fossil fuels and nuclear power (including guarantees for nuclear waste disposal from new power plants, loan guarantees, and subsidized insurance).
3. Eliminate subsidies for biofuels from food crops.
4. Build demonstration plants for key supply technologies, including central station solar thermal with heat storage, large- and intermediate-scale solar photovoltaics, and CO_2 capture in microalgae for liquid fuel production (and production of a high solar energy capture aquatic plants, for instance in wetlands constructed at municipal wastewater systems).
5. Leverage federal, state and local purchasing power to create markets for critical advanced technologies, including plug-in hybrids.
6. Ban new coal-fired power plants that do not have carbon storage.
7. Enact at the federal level high efficiency standards for appliances.
8. Enact stringent building efficiency standards at the state and local levels, with federal incentives to adopt them.
9. Enact stringent efficiency standards for vehicles and make plug-in hybrids the standard U.S. government vehicle by 2015.
10. Put in place federal contracting procedures to reward early adopters of CO_2 reductions.
11. Adopt vigorous research, development, and pilot plant construction programs for technologies that could accelerate the elimination of CO_2, such as direct electrolytic hydrogen production, solar hydrogen production (photolytic, photoelectrochemical, and other approaches), hot rock geothermal power, and integrated gasification combined cycle plants using biomass with a capacity to sequester the CO_2.
12. Establish a standing committee on Energy and Climate under the U.S. Environmental Protection Agency's Science Advisory Board.

AFTERWORD

by Dr. Helen Caldicott

The climate crisis has put the Earth in the intensive care unit. In the past few years I have experienced an acute sense of urgency to do my part to set it on the road to recovery. I have not felt such urgency since the threat of nuclear war between United States and the Soviet Union hung over the planet in the early 1980s, a threat incidentally that has not diminished, with thousands of Russian and US .nuclear warheads still on high alert, ready to be fired in minutes.

The Nuclear Policy Research Institute sponsored an energy conference in 2006 to which I invited some of the world's most experienced and able people in the energy field to ascertain whether they shared my sense of urgency about the state of the planet. This two day discussion dissected out the ecological and medical dangers of a fossil-fueled, nuclear-fueled energy system and explored the possibilities of a vibrant renewable energy economy.

Among the speakers were S. David Freeman and Arjun Makhijani. David's speech was extraordinarily inspiring as he raised the distinct possibility that all energy could be obtained from present-day technology without the use of fossil fuel or nuclear power. I could hardly believe my ears. This was an entirely new scenario that had never before been seriously entertained.

Dr. Makhijani agreed that the world was facing an ecological crisis and that the scale of the problem was escalating rapidly as grim news about climate alterations continued unabated. But was a renewable energy policy technically and economy feasible without nuclear power?

Arjun, one of the most capable scientists in environmental work, did not want to advocate something that he thought would only be feasible at an unbearably high cost. In his view, cost was part of the feasibility equation.

Several months of discussions took place before a plan of action eventuated. We agreed to initiate a comprehensive in-depth study to examine these questions. Dave Freeman and I would serve on an Advisory Board, along with other members from academia, industry, and the economic justice movement. To enable Arjun to focus entirely on the study, I agreed to accept the task of fundraising.

Arjun and I had many arguments as we discussed the conflicting goals which entailed urgency on the one hand and feasibility on the other. I reminded him that the patient had to survive at all costs and that from a medical perspective, the economy was secondary. He insisted that if we pushed things beyond what was economically feasible even with sensible policies, we would achieve nothing. We were not the captains of industry. We did not have our hands on serious capital to invest to help save the planet. The plan had to be within the realm of economic reality. It should frankly assess the current state of the technologies that were needed, how close they were to economical reality, and how these existing technologies could be marketed. We also needed a backup strategy if the main approach could not yield desired results.

The Roadmap meets all these requirements. Arjun has produced a study which fulfills my greatest hopes – an urgent action plan to move the Earth in a dignified way out of intensive care. This is a benign and efficient proposal to save the planet without the cancerous, radioactive, proliferation-prone side effects which current energy policy will inevitably bestow upon future generations. My message to all members of society, including local legislators, captains of industry, members of Congress, and presidential candidates is simple: read this book and act upon it.

Helen Caldicott, M.D.
Founding President, Nuclear Policy Research Institute

GLOSSARY

Baseload generation: Electricity generation on a continuous basis by large-scale power plants.

Biofuel: Fuel derived from biomass.

Biomass: Organic material produced by photosynthesis.

Cap: A limit on emissions.

Capacitors: Devices that store electric charge.

Carbon capture: Capture of carbon dioxide when fuels containing carbon are burned for their energy.

Carbon sequestration: Deep geologic storage of carbon for long periods (thousands of years) to prevent it from entering the atmosphere.

CFL: Compact fluorescent lamp, which is a high-efficiency light bulb.

CHP: Combined heat and power. In this arrangement, some of the energy derived from burning a fuel is used as heat (as for instance in heating buildings or for industrial processes), and some is used for generating electricity.

Combined cycle power plant: Power plant in which the hot gases from the burning of a fuel (usually natural gas) are used to run a gas turbine for generating electricity. The exhaust gas from the turbine is still hot and is used to make steam, which is used to drive a steam turbine, which in turn drives an electric generator.

Distributed grid: An electricity grid that combines significant portions of small-scale and intermediate-scale generation with centralized generation.

Earth-source heat pump: See geothermal heat pump.

Electrolytic hydrogen production: The use of electricity to separate the hydrogen and oxygen in water.

Geothermal heat pump: A heat pump that uses the relatively constant temperature a few feet below the earth's surface in order to increase the efficiency of the heat pump.

Ground source heat pump: See geothermal heat pump.

Hard cap: An absolute limit on annual emissions.

HFCs: Halogenated fluorocarbons. Could also apply to partially halogenated compounds.

IGCC: Integrated Gasification Combined Cycle plant. This plant gasifies coal or biomass and then uses the gases in a combined cycle power plant.

LEED: Leadership in Energy and Environmental Design – a rating system used for building efficiency. The platinum level is the highest rating.

Microalgae: Tiny algae that grow in a variety of environments, including salty water.

Nanocapacitor: Capacitors made using a nanotechnology that can store a very large amount of charge per unit volume. This technology is still in the laboratory stage.

Photoelectrochemical hydrogen: Hydrogen produced directly using devices similar to some solar photovoltaic cells that generate electricity. In this arrangement, hydrogen is produced instead of electricity.

Photolytic hydrogen: Hydrogen produced by plants, for instance, algae, in the presence of sunlight.

Pumped storage: Using electricity at off-peak times to pump water into a reservoir and then using a hydroelectric power plant to generate electricity with the stored water during peak times (or, when used with wind energy, when the wind is not blowing).

Smart grid: A distributed electricity grid in which electricity supply, electricity storage, and thermal storage (heat and coldness) are integrated with time-of-use controls of end-use equipment. It would enable real-time management of the electricity system so as to match electricity demand with the supply of intermittent renewable energy sources and reduce the total investment needed for a given level of energy services and reliability.

Solar light pipe: A fiber optic cable that conveys light from the sun along its length without leaking it out of the sides, much like a wire carries electricity. It can be used to light the interiors of buildings during the daytime.

Solar PV: Solar photovoltaic cells: Devices that turn incident sunlight into electricity.

Solar thermal power plant: A power plant that uses reflectors to concentrate solar energy and heat liquids that are then used to produce steam and generate electricity.

Spinning reserve: The capacity of electric power plants that are kept switched on ("spinning") but idle in order to be able to meet sudden increases in electricity demand.

Standby capacity: Power plants that are kept on standby to meet increases in electric demand.

Supercapacitors: See nanocapacitors.

Time-of-use rates: Electricity rates that vary according to the time-of-use relative to the availability supply and the types of electricity supply.

Ultracapacitor: A capacitor that can store much more electricity per unit volume than normal capacitors.

UNFCCC: United Nations Framework Conversion on Climate Change

V2G: Vehicle to grid system. Parked cars are connected to the grid. When the charge on the batteries is low, the grid recharges them. When the charge is sufficient and the grid requires electricity, a signal from the grid enables the battery to supply electricity to the grid.

APPENDIX A: NUCLEAR POWER

Uranium enrichment and reprocessing, once terms reserved for eggheads dealing in nuclear esoterica, are in the headlines everyday. Politicians and diplomats argue about them and the proliferation threats arising from the spread of commercial nuclear power technology.[1]

Yet, strangely, in a parallel universe also being played out on the public stage, fans of nuclear energy are proclaiming a "nuclear renaissance." The nuclear industry's claim, amplified by the megaphones of the media, is that nuclear power can play a vital role in saving the Earth from another peril – severe climate disruption caused by the anthropogenic emissions of greenhouse gases, particularly CO_2.

Could nuclear power really help save the world from what could be the worst environmental scourge ever to confront humanity? History would suggest two things: caution about the nuclear industry's messianic proclamations and careful analysis of the problem.

A. History

The early promises of the fervent advocates of nuclear energy were of an economic paradise that nuclear energy would usher in for everyone from the needy to the greedy. No whim or need would go unfulfilled. But it was mainly fantasy and propaganda.

Studies of the 1940s and 1950s showed that the public proclamations that nuclear power would soon to "too cheap to meter" were known then to be wrong. For instance, a 1950 article written by Ward Davidson, a research engineer with Consolidated Edison Company of New York, published in an industry journal, *Atomics*, concluded that the technical problems facing nuclear power were daunting. For example, the materials requirements would be stringent, given the high temperatures and damage from high neutron fluxes. Testing of the alloys to ensure the quality and uniformity needed would be difficult. All this meant, of course, that nuclear power would be expensive.

"Too cheap to meter" was part self-delusion, as shown by the florid and fantastic statements made by the most serious people, such as Glenn Seaborg, who led the team that first isolated plutonium, and Robert Hutchins, the President of the University of Chicago during the Manhattan Project. And it was in part organized propaganda designed to hide the horror of the hydrogen bomb. The statement itself was made in 1954, by the then-Chairman of the U.S. Atomic Energy Commission, Lewis Strauss. It was part of a campaign to convince the world that the American atom was a peaceful one. There was fear that the Soviets would do that first.

In September of 1953, less than a month after the detonation of the Soviet's first hydrogen bomb, AEC Commissioner Thomas Murray wrote to the commission's chairman that the U.S. could derive "propaganda capital" from a publicity campaign surrounding their recent decision to construct the Shippingport nuclear power plant.[2] Sterling Cole, the chairman of the Joint Committee on Atomic Energy in the U.S. Congress, reached a similar conclusion regarding the importance of demonstrating the "benefits" of nuclear power as a counterbalance to the immense destructive force of the hydrogen bomb. This conclusion, in fact, led Cole to worry that the Soviets might beat the U.S. to a functional nuclear power plant, and thus steal the claim to being the true promoters of the "peaceful" atom. In a letter to a fellow Congressman, Sterling Cole wrote

> It is possible that the relations of the United States with every other country in the world could be seriously damaged if Russia were to build an atomic power plant for peacetime use ahead of us. The possibility that Russia might actually demonstrate her "peaceful" intentions in the field of atomic energy while we are still concentrating on atomic weapons could be a major blow to our position in the world.[3]

As early as 1948, the Atomic Energy Commission reported to Congress that "the cost of a nuclear-fuel power plant will be substantially greater than that of a coal-burning plant of similar capacity."[4] In the January 1949 issue of *Science*, Robert Bacher, one of the original members of the AEC and a member of the scientific team at Los Alamos during World War II, cautioned that despite the progress that was being made, it was "far too early to make any predictions about the economic feasibility of atomic power."[5]

One of the most direct of the early critiques of the economics of nuclear power came in a December 1950 speech before the American Association for the Advancement of Science by C.G. Suits. At the time, Suits was the Vice-President and Director of Research at General Electric which was then operating the Hanford plutonium production reactors in Washington State and was one of the principal companies developing nuclear reactors for the production of electricity. In his speech, which was reprinted in the industry journal *Nucleonics*, Suits stated bluntly that:

It is safe to say… that atomic power is not the means by which man will for the first time emancipate himself economically, whatever that may mean; or forever throw off his mantle of toil, whatever that may mean. Loud guffaws could be heard from some of the laboratories working on this problem if anyone should in an unfortunate moment refer to the atom as the means for throwing off man's mantle of toil. It is certainly not that!

… At present, atomic power presents an exceptionally costly and inconvenient means of obtaining energy which can be extracted more economically from conventional fuels… The economics of atomic power are not attractive at present, nor are they likely to be for a long time in the future. This is expensive power, not cheap power as the public has been led to believe.[6]

In 1953, an official AEC study concluded that "no reactor could be constructed in the very near future which would be economic on the basis of power generation alone." Significantly, this language was identical to that in a study published by industrial companies and major utilities including Bechtel, Monsanto, Dow Chemical, Pacific Gas and Electric, Detroit Edison, and Commonwealth Edison.[7]

The dismal assessment of the prospects of nuclear went back to the Manhattan Project. In a star-studded 1948 report, authored by Enrico Fermi, Glenn Seaborg, and J. Robert Oppenheimer, the authors concluded that there was "unwarranted optimism" about the speed with which the technical difficulties facing nuclear power could be overcome. Ironically, the self-same Glenn Seaborg waxed eloquent about how plutonium fuel could transport everyone into a technical wonderland of "planetary engineering" – which, of course, could only be done if energy were actually very cheap.

A large part of the idea that nuclear energy would be a wondrous energy source was based on the idea that some kinds of nuclear reactors, called breeder reactors, could make more fuel than they consumed. Uraniun-238, not a reactor fuel, would be turned into fuel in breeder reactors, even as those same reactors consumed plutonium fuel. The net result would be more fuel at the end of the cycle. Since uranium-238 is a plentiful isotope in nature, the fantasy was only slightly exaggerated from a pure physics point of view.

But experience has shown that physics is not enough. An energy source must still meet the tests of safety, reliability, and cost. In the case of nuclear energy, there is also the unique problem of nuclear proliferation, in part hidden in the form of the plutonium content of the spent fuel and in part in the form of the spread of know-how. Taken together, these factors made the physics "magic" evaporate the first time around. Breeder reactors and the associated reprocessing have yet to be commercialized after over $100 billion in expenditures worldwide (constant 1996 dollars) and more than fifty years of effort. France, which has the most experience in the use of plutonium as an energy source as well as the largest commercial infrastructure for that purpose, has spent an extra 2 cents per kWh on electricity generation from plutonium fuel used in its light water reactors. The main breeder reactor that has been used in commercialization efforts has used liquid sodium as a coolant. It has a very mixed history, from reasonably good performance to utterly dismal. The largest such reactor, Superphénix, a 1,250 megawatt machine, was built in France. It operated for 14 years at an overall capacity factor of less than seven percent. Even if poor performance is discounted, breeder reactors remain far too expensive relative to light water reactors, the main design in use today. And since they would use plutonium (mixed with uranium) as the fuel, they pose greater proliferation risks.[8]

Half a century of efforts to commercialize thorium breeders – reactors that make fissile uranium-233 out of non-fissile thorium-232 – have not yielded a single commercial machine.

We have commented on some current proliferation issues in the preface and would not repeat that analysis here. But it is worth noting that the potential of nuclear power to provide a hidden infrastructure for nuclear weapons has long been known. In fact, that very possibility was entertained for the United States in 1946 by none other than J. Robert Oppenheimer, who was then the chairman of the General Advisory Committee of the Atomic Energy Commission. He did so in the context of the possibility that there would be a convention on international control of nuclear weapons that would result in nuclear disarmament:

> We know very well what we would do if we signed such a convention: we would not make atomic weapons, at least not to start with, but we would build enormous plants, and we would call them power plants – maybe they would produce power: we would design these plants in such a way that they could be converted with the maximum ease and the minimum time delay to the production of atomic weapons, saying, this is just in case somebody two-times us; we would stockpile uranium; we would keep as many of our developments secret as possible; we would locate our plants, not where they would do the most good for the production of power, but where they would do the most good for protection against enemy attack.[9]

Six decades later, quite a few countries may be taking a leaf from this book, or at least considering it.

B. Nuclear Waste

Even though efforts to commercialize plutonium have failed miserably, proposals to reprocess spent fuel, which contains about 1 percent plutonium (total content of all plutonium isotopes), are being revived. A central claim made now is that reprocessing will reduce the problem of disposal of spent fuel, which contains over 99 percent of the radioactivity associated with commercial nuclear power.

The vast majority of nuclear reactors in the world today are light water reactors, which use low-enriched uranium as a fuel. This fuel contains three to five percent uranium-235, which is the fissile isotope of uranium that can sustain a chain reaction. Almost all the rest is uranium-238. Once a reactor is fueled, U-235 atoms are bombarded with neutrons and they split, liberating energy and more neutrons. Some neutrons split more U-235 and some are absorbed in the more plentiful U-238, converting it into plutonium-239. Some of this plutonium fissions also yielding energy, and some remains until the fuel must be removed from the reactor. The typical composition of fresh fuel and spent fuel are shown in Table A-1.

Table A-1: Pressurized Water Reactor Fresh and Spent Fuel Composition (rounded)

Material	Fresh Fuel (weight percent)	Spent fuel (weight percent)	Comments
Uranium-235	4	1	Each kilogram of enriched fuel creates about seven kilograms of depleted uranium in the course of enrichment.
Uranium-238	96	94	
Plutonium (+ smaller amounts of other transuranic radionuclides)	0	1	Mixture of various isotopes from Pu-238 to Pu 242. Can be used to make nuclear weapons if separated from the rest of spent fuel. Predetonation is more likely for bombs made with reactor-grade plutonium than with weapon-grade plutonium.
Fission products	0	4	Fission products contain the vast majority of the radioactivity in the spent fuel.

Note: Trace quantities of U-234 and activation products are not shown.

In the early days of nuclear power, it was assumed that scarcity of uranium would lead to plutonium becoming the main fuel for nuclear power plants. But uranium was more plentiful than thought and reprocessing and plutonium fuel (which generally consists of mixed plutonium dioxide and uranium dioxide) turned out to be costly. The proliferation risks of reprocessing also became more clear after India exploded its "nuclear device" in 1974. Presidents Ford and Carter took steps to end the development of the plutonium economy in the United States. President Reagan tried to revive reprocessing in the early 1980s, but there were no commercial takers. To President Reagan's credit he did not propose massive new subsidies or that the U.S. government should enter the plutonium commercialization business.

In order to relieve utilities of the burden of spent fuel that now had no place to go and to reduce long-term proliferation risks arising from spent fuel sitting around at dozens of sites, a deep geologic repository program was created in 1982 pursuant to the Nuclear Waste Policy Act. Its history is a long and tangled one, but soon (1987) all resources were focused on just one site – Yucca

Mountain, Nevada. This is, in my opinion, the worst repository site that has been investigated in the United States. Indeed, the DOE's own assessment of the merits of the geologic setting in containing the radioactive waste, should they leak out of the containers, is that it would add almost nothing to the site's performance. Essentially the entire burden of performance, that is, keeping doses low enough to meet standards, would be on the packaging. Even so, the rules and standards have been changed numerous times, since Yucca Mountain has had serial difficulties in meeting proposed radiation exposure limits and engineering performance standards. For instance, Yucca Mountain was originally supposed to meet the 1989 EPA regulations that apply to all deep geologic repositories. Subsequently, the EPA Science Advisory Board found that Yucca Mountain may not meet the carbon-14 emissions standard.[10] Instead of looking for a new repository, Congress mandated that a new set of standards specific to Yucca Mountain should be created. A quarter century after the passage of the Nuclear Waste Policy Act, a new EPA standard for the Yucca Mountain repository has yet to be finalized.

As another example, the Nuclear Regulatory Commission published the criteria for performance of a geologic isolation system in 1985. These criteria placed primary emphasis on the properties of the geologic setting to prevent significant amounts of radionuclides from reaching the human environment. Only secondary reliance was placed on the disposal containers and associated engineered barriers in the containment of the pollution. But Yucca Mountain is made of a rock known as volcanic tuff, which turned out to be a poor candidate by these criteria, so the criteria were scrapped. The new criteria stressed "total system performance"; in effect, the performance criteria for the geologic system were scrapped. DOE's own estimates show that it is now placing primary reliance on the container. Moreover, the canisters are made of metal and their susceptibility to corrosion is highly variable, depending on the environmental conditions in the rock.[11]

Reprocessing only makes the problem worse, even though it is promoted as "recycling." The "recycling" portion generally applies to just that one percent of spent fuel that consists of plutonium isotopes. In the absence of economical breeder reactors (which still remain a nuclear pipe dream), the plutonium would be used as mixed oxide fuel in light water reactors at considerable expense. The current commercial reprocessing technology, PUREX (for plutonium-uranium-extraction) is huge and polluting. The largest such installation in the world is located on the Normandy peninsula in France. The radioactive liquid waste discharges from that and the similar facility in northwestern England, have polluted the seas all the way to the Arctic Ocean. Ten of the twelve parties to the Oslo-Paris accords (OSPAR) have asked the French and British to stop the discharge, but they have not done so. (The other two parties are France and Britain; they abstained and hence are not bound by the vote.)[12]

The fission product stream, which has most of the radioactivity, would still need to be disposed of in a deep geologic repository. Most of the long-lived radioactivity in this stream consists of cesium-137 and strontium-90, with half-lives of about 30 and 29 years respectively. But there are also significant amounts of iodine-129 and cesium-135, which have half-lives in the millions of years. While the volume of high-level waste is reduced after it is solidified in a glass matrix, reprocessing creates additional streams of waste besides the liquid discharges noted above. Specifically, intermediate-level waste, a waste classification used in France and other European countries, would be created in significant amounts. This waste must be disposed of in a geologic repository as well. Overall, reprocessing increases the volume of radioactive waste greatly when all waste streams are taken into account and does not eliminate the need for a deep geologic repository.[13]

The uranium stream that results from reprocessing consists of 95 percent of the nuclear material weight of spent fuel (U-238 plus U-235). It becomes contaminated with traces of fission products, notably technetium-99, as well as plutonium and neptunium-237. The contamination with these materials, which are much more radioactive than the uranium itself, creates considerable problems for the re-use of the uranium. Before it can be used again, it must be chemically processed and re-enriched to 3 to 5 percent U-235 content. The trace contamination results in contamination of the enrichment plant and creates additional radioactive exposure hazards for workers. For instance, in 1999, the Paducah uranium enrichment plant in Kentucky became notorious for not having warned its workers adequately about these trace contaminants in the uranium.[14] A subsequent analysis determined that plutonium and neptunium were concentrated in certain process streams at the plant and created the potential for high worker doses.[15] Trace contamination with plutonium and other radionuclides at Paducah was an important factor in the legislation that Congress passed in the year 2000, setting up a compensation program for nuclear weapons workers made sick by exposure to radiation and chemicals. The Paducah plant belongs to the U.S. Department of Energy; it is currently used only for commercial uranium enrichment. In the past it was used both for military and commercial purposes.[16]

While public information is scarce, it is interesting to note that France sends at least some of the contaminated uranium recovered at its La Hague reprocessing plant to Russia rather than re-enriching at home. If reprocessed uranium were to be disposed of as a waste instead of being re-enriched, this would also pose considerable problems. They would be more difficult than those faced by depleted uranium because the specific activity of the reprocessed uranium is roughly double that of depleted uranium; in addition it contains transuranic and fission product contaminants

Finally, all uranium enrichment results in a stream of depleted uranium, which is uranium depleted in the fissile isotope U-235. Depleted uranium consists mainly of the non-fissile isotope uranium-238 (99.7 to 99.8 percent usually). Some of this depleted uranium has been used for a variety of commercial and military purposes, the latter including tank armor and shells that have spread contamination on battlefields and testing areas in several countries. But the vast majority of it still remains as an orphan waste of the commercial and military nuclear enterprise. There is at present no place to dispose of depleted uranium in a way that would conform to prevailing radiological safety and health norms. Nor is there any program in place find one. It will not be easy. The characteristics of the waste make it akin to what is called transuranic waste (or Greater than Class C waste) and it should be handled accordingly – that is disposed of in a deep geologic repository.[17] But the depleted uranium sits at various sites in nuclear states, including three in the United States – Oak Ridge, Tennessee, Paducah, Kentucky, and Portsmouth, Ohio.

The Global Nuclear Energy Partnership

Based on a U.S. initiative, the United States and Russian governments launched a collaborative effort in 2006, called the Global Nuclear Energy Partnership (GNEP).[18] According to this proposal, countries that currently have reprocessing or uranium enrichment capacity would be allowed to possess it and, if they wish, expand it, while those that do not, would be prohibited from acquiring it. In return, the reprocessing-enrichment haves would supply the have-nots with prepackaged reactors and fuel. The spent fuel would be returned to the haves.

GNEP is a transparent attempt to rewrite Article IV of the NPT, which guarantees an "inalienable right" to acquire commercial nuclear technology to the non-nuclear weapons states that are parties to it. The inclusion of Article IV was unfortunate, but it was a fundamental part of the bargain. Nuclear energy had been romanticized and politicized at least since President Eisenhower's famous "Atoms for Peace" speech at the United Nations in December 1953. Article IV was a direct descendant of the U.S. Atoms for Peace program that followed that speech.

The second part of the NPT bargain was that nuclear weapons states would eliminate their nuclear arsenals.[19] The latter now recedes into the far future – all five nuclear weapon states parties to it are modernizing their arsenals. What India used to call "nuclear apartheid" before it detonated its own nuclear weapons in 1998, is being perpetuated. GNEP would extend that to nuclear energy. There are unlikely to be any serious takers. On the contrary, more and more countries are expressing interest in nuclear power, with the not too hidden agenda of acquiring much of the nuclear infrastructure and most of the knowledge that would enable them to make nuclear weapons in the future. There is even an active debate in Japan today about whether it should become a nuclear weapon state. Should it decide to do so, its reprocessing capability, its stocks of commercial plutonium, and other technological infrastructure would probably enable it to become a nuclear weapon state in six months.[20]

There are other potential components in GNEP, including a reprocessing technology called "electrometallurgical processing." Despite the fact that it would not separate pure plutonium, it would create material that non-nuclear states or terrorist groups could use to make nuclear bombs. Moreover, being more compact than PUREX, it would be far easier to hide the separation facilities, making them more proliferation prone, not less.

The costs of GNEP are likely to be huge. GNEP is not going to solve the problem of nuclear waste. However, it may be a new source of funds for that part of the nuclear power establishment that is closest to the weapons bureaucracy or is part of it. GNEP is centered in the Department of Energy, which owns the nuclear weapons complex.

There is no really good solution to the problem of spent fuel and high-level waste disposal. It is very difficult to compute the impacts on generations far into the future. Is it sensible to go on creating wastes that risks contamination of water, with its attendant radiation health damage, far into the future? Yet the problem of leaving it on the surface indefinitely is even more difficult. It entails the risks of proliferation (via reprocessing), terrorism, and accidents. Hardened On-Site Storage of spent fuel – that is, storage that could withstand severe attacks without dispersal of huge amounts of radioactivity – for a few decades followed by disposal in a deep geologic repository are the "least bad solution." But that "solution" makes sense only if we limit the creation of waste in the future.

C. Cost

The history of cost overruns at nuclear plants in the United States is well known.[21] Significantly, in a review of historical experience with nuclear plant construction, the DOE's Energy Information Administration noted explicitly that

> ... although the utilities did increase their lead-time and cost estimates as work on the plants proceeded, they still tended to underestimate real overnight costs (i.e., quantities of land, labor, material, and equipment) and lead-times even when the plants were 90 percent complete.[22]

> In this review, the Energy Information Administration found that, for those plants that began construction between 1966 and 1969, the utilities were underestimating the final cost of the nuclear plants by an average of 63 percent prior to construction beginning and were still underestimating their final cost by 22 percent when the plants were three-quarters complete. Surprisingly, for those plants that began construction between 1974 and 1977, the nuclear industry actually grew slightly worse at estimating the final plant cost despite its increase in experience. Specifically, the utilities underestimated the costs of these plants by 72 percent prior to construction and, even when past plants were three-fourths complete, they were still underestimating the final construction cost by roughly 23 percent.[23]

One reactor that is being commonly considered in cost studies is Westinghouse's AP-1000.[24] An AP-1000 has never been built anywhere in the world, not to mention anywhere in the United States, so no real world experience is available from which to draw a direct comparison. While it is the same overall concept as the pressurized water reactor, the many new design features, some added for safety, add to the uncertainty in cost estimates.[25] As noted by analysts at Standard & Poor's in their 2006 assessment of nuclear power generally, "given that construction would entail using new designs and technology, cost overruns are highly probable."[26]

In recent regulatory actions in North Carolina, where Duke Power has proposed to build new coal plants at the existing Cliffside power plant, the doubts about nuclear power's cost-effectiveness and viability were voiced. Jim Rogers, CEO of Duke Power, which has expressed serious interest in pursuing nuclear power stated in his testimony:

> Here's my judgment. We put 1800 [dollars] in because it's what Westinghouse has told us the number is. We are in negotiations with Westinghouse. My personal – and we modeled – what if it was 2200 and under 2200 Cliffside and Gas would be the least cost alternative in every scenario almost. And the reality is, my personal belief about nuclear, I don't think it comes on in 2016. I'm not a true believer. And secondly, I don't believe – I believe it comes closer to 2500 or 2600. And if you look at the testimony of Judah Rose, it's pretty close to 2500. So my personal judgment is, is that nuclear comes in at a much higher price, and it comes – and we are actually able to build it, it's going to be delayed beyond 2016. That would be my bet if I had to make the bet today.[27]

Coming from the CEO of Duke Power, this is an especially interesting statement. Duke Power is a member of the U.S. Climate Action Partnership (US-CAP) of some corporations and large environmental groups that has endorsed the concept of a 60 to 80 percent reduction in U.S. greenhouse gas emissions by 2050.

The U.S. Congress is considering ever more massive subsidies for nuclear power plants in the form of loan guarantees – possibly as much as $4 billion to $5 billion per reactor for as many as 28 reactors.[28] The reason is clear: the economic risks of nuclear power plants are just too large. In the words of Michael J. Wallace, who co-heads UniStar Nuclear, a company that wants to build nuclear power plants: "Without loan guarantees we will not build nuclear power plants."[29] We have already noted the opinion of a leading credit rating company, Standard & Poor's, that the credit standing of a company ordering a nuclear power plant would weaken if it ordered a nuclear power plant, even if it did so with government support (see Chapter 7).

D. Nuclear Power and Global Climate Change

There are two schools of thought among proponents of nuclear power and climate change. One is that a large number of reactors would be built to reduce the need for more coal-fired power plants. The other school advocates that nuclear power should be kept in the mix since all available energy sources that could help reduce CO_2 emissions should be considered as options.

If nuclear power is used as a principal element of future electricity generation worldwide, a very large number of reactors would have to be built in the coming decades. Brice Smith has estimated that for nuclear power to contribute about 20 percent of the global electricity supply by mid-century, about 1,000 reactors of 1,000 megawatts each would have to be built. For nuclear power to play a role comparable to coal today – about half of total generation – 2,500 reactors would have to be built in the same time. This is a rate of one reactor every six days.[30]

Such a massive system would require a new repository every few years, two or three new enrichment plants every year. It would greatly increase pressures for reprocessing. The risks of accidents would increase, even disregarding the potential for sloppy construction if the number of reactors is increased rapidly. Brice Smith has estimated that if 2,500 reactors are actually built in forty years, there would be a sixty percent chance of a Three-Mile-Island type of meltdown even if the safety of reactors were increased by a factor of ten compared to the present.[31]

But even far less serious events can trigger doubts about the nuclear industry as a whole, making it an unstable way to plan for future electricity generation. The July 16, 2007 earthquake in Japan under Tokyo Electric Power Company's 8,000 megawatt, seven-reactor nuclear power plant is a case in point. The leak of radioactivity into the sea was not large. *Nature*, a journal of science that has editorialized on nuclear power, noted its vulnerabilities after the earthquake and the poor public communications by Tokyo Electric that followed:

> Global warming and high energy prices have put nuclear power firmly back in the picture around the world. Plans are afoot to build new plants in Britain and the United States, and China and India look set to press ahead with nuclear power on a significant scale.
>
> Investors in planned nuclear plants continue to worry about waste disposal and liability issues, and look to sympathetic governments to provide assurance regarding these. Lurking in the back of their minds, however, is the ever-present risk of accidents of the sort that played havoc with the global industry at Three Mile Island, Pennsylvania, in 1979 and at Chernobyl in 1986. Another such event could undermine political support for nuclear power and so up-end their planned investments altogether, possibly before a single megawatt of power is generated and sold.[32]

How much can one rely on an energy source whose acceptability may depend on whether there is a severe earthquake or accident somewhere in the world and on the care with which geologic faults have been studied and incorporated into the design? Nuclear power is unique in having this vulnerability. No coal mine accident, oil tanker spill, or natural gas explosion puts the whole industry into question. Only climate change, which is being created by the global use of fossil fuels, has done that. But the nuclear industry could be derailed by a single local event – a severe accident, or possibly even by a single earthquake, to say nothing of a serious terrorist attack. More power plants would simply multiply the risks. Finally, the heat waves and regional droughts that are likely to accompany rising global temperatures threaten to make nuclear power into an intermittent source in the summers. For instance, one of the three nuclear reactors at Browns Ferry, belonging to the Tennessee Valley Authority, was temporarily shut down in August 2007 because the river water used to cool it was too hot. Sufficient cooling water was available for only two of the reactors.[33] Similar problems were experienced in France in 2006 when reactor power output was reduced[34] and in 2003.[35]

Those who have advocated that nuclear power should be kept in the mix have not really addressed the risks of doing so versus the option of simply omitting it from the energy picture and creating a reliable grid without it.

APPENDIX B: INTERVIEW REGARDING INDUSTRIAL GREENHOUSE GAS EMISSIONS

Final Summary of telephone conversation with Dawn Rittenhouse and John Carberry, both of DuPont, with Arjun Makhijani, 14 February 2007.

Reviewed and corrected by Dawn and John. Edits accepted and document cleaned up by Arjun. Notes are not verbatim, but a summary that reflects the substance of the conversation.

1.*What procedures do you have for GHG [greenhouse gas] emissions accounting in DuPont? Are there plant level measurements and reporting procedures so that HQ can compile company -wide data?*

Dawn and John: We use WRI's [World Resource Institute's] GHG protocol to calculate emissions. We use a control approach – that is accounting for 100% emissions of operations over which we have control. Scope 1 accounts are associated with direct use of fuel; Scope 2 is purchased electricity and steam. We don't do supply-chain-related emissions. Our corporate plan includes all environmental goals. Each site in May and June enters information into that plan and then it is pulled together at the corporate level to provide the overview.

Arjun: So basically you account for GHG emissions from fuel and energy purchases?

John: Yes. We don't do personal commuting and business travel. I did a check once and found that it would not change things significantly to include this. It would be 3 or 4 percent increase. However, for some businesses, like pharmaceuticals, travel by employees can be large.

Dawn: We have a new goal on our marketing fleet. It's not a big fleet. We are working with PHH, who is our fleet provider, and Environmental Defense, to do calculations on GHG emissions of our fleet. So we are reducing GHG based on using leading technology. That is associated with our fleet goal for 2015 – all of our fleet will be using a leading technology by that date. We did not define what that technology would be.

As for GHG emissions, our plan is to reduce emissions a further 15% off the 2004 base.

2. Do large plants have energy managers whose responsibilities include ensuring that decisions such as replacement of motors and lighting are made with energy efficiency in mind?

John and Dawn: We have a corporate energy competence center network – it is a formal network of energy professionals around the company. It is their responsibility to implement the energy efficiency programs of the company. The network is to share expertise and learning. If you recall the Pew Case study – there was a write up on the energy leader for the titanium technologies business, I sent you – that is an example. Craig Heinrich leads the energy work for the titanium technologies.

Each manager at the larger sites (20 plus largest sites) has goals and targets. And they go after those by addressing a wide range things. Their efforts are not limited to lighting and motors, but also go to areas like steam management and process changes.

We are committed to corporate leadership for manufacturing excellence. There is a corporate leadership process for manufacturing excellence. Energy efficiency is one of their top priority initiatives. They assign energy efficiency goals, monitor the progress of the site energy managers, and provide assistance where appropriate.

Arjun: Is a one or two percent [energy use] reduction per year reasonable as an energy efficiency goal across industry?

John: On an absolute basis, yes. Not if it is indexed to GNP. 1-2% in excess of GNP growth rate will probably be needed.

Probably the most important thing is to recognize that the four segments of the energy economy – residential, commercial, transportation, and industrial – have different marginal prices for energy. There have been different arguments about how to control different sectors. You have to take into account the differences between the sectors. But any one of the approaches would be suitable provided that it translates effectively into an effective market mechanism and gives credit for early action.

Arjun: I think cap and trade may be better in industry than standards, which I think would be better for appliances and the residential sector. What do you think?

John: The EU is experimenting with efficiency standards and cap and trade. The Dutch are making a good run at efficiency standards for industry. Big problem with cap and trade is adjusting for the evolution of industry.

Arjun: I propose that there be caps for an entire sector or industry segment and auction off the allowances on the market. Then the cap can be reduced every one to two years. That way you automatically get credit for early action because you don't have to buy so many allowances. An extra benefit could be given to companies that take early action by giving extra points in the federal contracting score for lower GHG emissions per dollar, for example.

Dawn: Your cap and trade proposals are way too complex in their details to comment on, briefly here. In short, an economic driver by industry sector, and as broad an application to that sector as reasonable, should be guiding principle.

John: I agree with concept of government leading the development of a market and taking into account the efficiency or GHG emissions of the suppliers. But there is no reason why companies should not also lead in the same way. They too could select suppliers based on their GHG performance and in fact some companies already do this.

Arjun: The concept of capping a segment of industry would be to limit it to large industry. I am leaning to a hybrid approach with cap and trade for larger users of fossil fuels and efficiency standards for smaller users, for instance in the residential and small business sector. The paperwork would be too much for small business.

John: This is a sore point for large businesses as well. No one wants paperwork – it is a burden on all. But I agree that large business can be more efficient in doing that paperwork. If there are ten major paperwork requirements, then in a large company each single requirement can be done by one individual, totaling ten people because there is enough work in each area. In a small company, the same person has to do all of them and so specialization is not possible. Although there is some efficiency gain, the cost (per unit of sales) is still about the same.

3. *What are the main areas in which DuPont seeks to achieve its reduction in energy use from its 2004 base? I am not looking for specific numbers and plans, but the areas of priority according to economic opportunity and to get some sense as to whether the same may apply to the rest of the chemical/biotechnology industry.*

Dawn: I want to make sure that we get clear as to what DuPont is doing. In 1999, we had a goal to reduce GHG emissions by 65%. By 2003 we had reduced by 72%. In 2004 we sold off our fibers unit- Invista, which produced nylon, PET and Lycra fibers etc. So we re-baselined the GHG goals to 2004 so there would not be GHG accounting problems due to the sale. We will reduce GHG emissions by an additional 15% by 2015 compared to 2004. The targeted areas are HFCs as well as energy projects.

We also set a goal in 1999 of holding energy use flat, based on the 1990 level. We were actually 6% lower than that in 2005 which is the last year we have the data collected. Then we reset the base to 2004, no we did not reset that base line- we just subtracted the energy from the 1990 number that was associated with Invista so that we can still use the 1990 as the baseline so it would be based on energy consumed by companies we are actually operating. We continue to monitor that. We continue to work on energy efficiency. On top of that we have a goal of getting 10% energy from renewable sources.

There is no single answer, nor even just a few. In broad terms major improvements come from:
* Improvements in first-pass, first-quality yield
* Maximization of process through-put and process up-time
* Combined heat and power generation (CHP)
* Changes to processing equipment
* Improved process control
* Powerhouse generation and distribution systems (steam traps, insulation, motor efficiency, etc.)

4. In the USCAP paper, the coalition recommended a goal of reducing GHG emissions by 60 to 80% by 2050. Is DuPont or USCAP developing active plans for the 60 to 80 percent reductions in GHG emissions? What priority areas of research should the federal government undertake that would help achieve that goal?

Dawn: That goal for 60-80% is for 2050 and it is not for one company. We are talking about expectations of energy efficiency and new forms of economically efficient energy supply, as well wind and solar energy. Through the next 45 years that will allow us to get to that goal. USCAP did not get into any kind of detail as to how one would get there [to a 60 to 80% GHG emissions reduction]. This is a man-on-the-moon type of thing – set a big goal and get people focused on meeting the big goal.

John: OK, what is that 60 to 80% reduction going to look like? If you got it down to a specific level – for instance, it would be how much energy is used in housing or different sectors? But in all cases, it is going to be the sum of a huge number of things that will need to be done. There is an overarching set of things-say renewable or passively safe nuclear energy or clean coal – those have applications across all segments. But as soon as you say a specific industry like the chemical industry, you have a lot of details that go into it – alternatives to distillation, for instance. Then there are the other GHGs besides CO_2 – capture of methane from offshore platforms, coal mines, and landfills. If you got 20 or 30 techies spread broadly across disciplines into a conversation, you would get 200 to 300 good ideas. There is the green building initiative. They have a whole bunch of things on HVAC design and passive heating and cooling. "There ain't no silver bullet and we don't want any lone rangers," as one of our engineers says. I could come up with 50 items if you gave me an hour. Take just the chemical industry or a segment of that, you'd have a host of specific things. The answers would be markedly different than in the aluminum industry. We'd have the big four [supply options] across the board – passively safe nuclear with acceptable waste management, clean coal with CO_2 sequestration, environmentally sound biomass, and reliable wind power with real solutions for managing the storage and distribution.

In the chemical industry CHP [combined heat and power] is a big one.

Another is replacing distillation – one alternative is modernization of processes so you don't have so many operations that involve distillation. Or it could be replaced by crystallization or membrane separation technologies, for example. Other areas are steam system management, insulation, powerhouse modernization, steam trap management. Optimization for first pass first quality yield is a big one – that is, make it correctly the first time. If you don't make it correctly, you have to recycle the product and make it again and you have waste all the energy that was used the first time.

Optimizing the manufacturing efficiency of your facility is another one. If you are in a standby hot mode, you use 60 or 70% of the energy anyway. So you want to run 100% of capacity 100% of the time. Then there is optimized process control and finding alternatives to grinding of solid materials – grinding is highly energy intensive.

The kind of question you are asking how are you going to get there [to 60 to 80 percent reductions], I probably would have to have a list of 10 to 20 big ones if I could get together the technical people from various areas.

One thing that we could have mentioned is the work on industrial biotech – for instance, the production of PDO from a bio route versus a chemical route is allowing us to save considerable energy – LCA [life-cycle analysis] results demonstrate that Bio-PDO™ requires 40-50 % less total energy to make that chemically derived PDO [polyester monomer propanediol].

Arjun: How about more waste heat recovery? It seems to me that developing new heat exchange materials that allow for more efficient transfer across small temperature differences – a few tens of degrees – would be helpful.

John : Improved waste heat recovery could come in at least two ways:

- Significant improvements in creating heat transfer surface area without an excessive capital cost or pressure drop penalty
- Significant increase in the minimum operating temperature for equipment that converts waste heat to electrical energy without an excessive capital cost or pressure drop penalty, or some other operational problem such as sensitivity to corrosion or fouling.

Arjun: How about using nanotechnology to increase the heat transfer surface area? They are attempting that in nanocapacitors to increase storage of charge per unit weight dramatically.

John: I have not seen anything that will say nanotechnology will give a big area without a bigger pressure drop. This a large dynamic world that is very complex that is set up that allows for innovation. I will go along with a cheap way to get a lot bigger area. That would make a lower delta-T [temperature difference] practical.

5. What part of steam generation is done by combined heat and power and what part by boilers alone? In other words, is there a large or small scope for DuPont to increase efficiency by going to CHP?

John: Most large manufacturing facilities already utilize CHP either onsite or through purchase of steam (and electricity where permitted) from a third party that owns the CHP facility itself. Some additional potential CHP capability exists, but current energy, electricity, and equipment prices are such that economic justification is difficult.

6. Has DuPont considered going to CHP plus carbon capture in algae and then production of fuels from algae. This system has been developed at MIT and used in their 20 MW CHP. See http://www.greenfuelonline.com/technology.htm. How would you rate this system compared to your recent biobutanol project?

John and Dawn: I am not aware of us doing anything in that specific area. The whole point is – let's make sure we promote those technologies that convert biomass into high value products. Where the biomass comes from – there are a lot of options for that. The two issues are not necessarily connected. Algae farming has been mentioned as a source of carbon. Others emphasize maximizing carbon capture in the farming industry. For instance, they burn rice hulls in the open air today. There are lots of potential sources of carbon – food industry, animal farming, algae – you've go to create a world that permits the best of them to emerge, There could be algae farming in the Gulf Mexico, but there are environmental arguments against it.

But if algae farming became a big industry, DuPont would probably be interested in it. And DuPont would be interested not only for fuel but also up the value chain.

7. Can DuPont's petroleum and natural gas feedstock requirements be met nearly fully with biomass-derived hydrocarbons?

Dawn: The question is not whether DuPont can meet its own requirements that way. We are working to get the raw materials that we need from biomass. The question is as a whole society can and if we do that, whether we will have any ecosystems left.

Arjun: I think that ethanol from food is not a good idea – turning fuel into food and food back into fuel is going to be inefficient especially when the solar energy capture is low. Biofuels have to be done much more efficiently.

John: I agree that the idea that you are going to grow wheat for methane is not good. First maximize the carbon capture rate of the farm and make the maximum use of the highest value carbon. Then collect the waste carbon for fuel but in a way that we don't deplete the soil. The grain can be used for food and the residues for other things.

John: We have looked at the question of feedstocks from biomass for DuPont some. There is enough for DuPont – but that is not the issue – because DuPont is not the only one competing for it. The power industry is willing to pay a higher price for natural gas than the chemical industry. Their supply and demand is here, but we have to compete with lower price of natural gas in other countries. There is a huge difference in that issue again. Presently, the molecular structure of biomass carbon is not quite right for many applications. Or we have to come up with alternative products. But Mother Nature doesn't give us [the chemical industry] the exactly correct molecules. We have to learn to adapt our supply needs to what is provided, and to modify what nature provides.

Dawn: But if you think back to the corn biorefinery, our goal is to get the raw materials from that.

John: Yes, that is the ultimate goal. It is a matter of timescale and costs.

8. *What are the current prevailing industry assumptions about the price of natural gas and of the cost of hydrogen derived from it? I am not asking for DuPont-specific assumptions, which I presume would be confidential, but for your sense of the general thinking in the industry about natural gas prices and hydrogen costs.*

John: Steam reforming of methane is the preferred way of coming up with H. This is used for hydrogenating chemicals, but we could not use it as fuel.

What is the price of hydrogen for this high value H? We will need to contact Air Products. It is higher than the fuel value.

Let me go to the making H – using H as form of energy storage. Make sure that you have properly considered capital cost. You have a large amount of capital that you are using only part of the time. So electrolysis you only use for a third of the day. So your capital effectiveness factor is only 0.3, not counting anything else. Then there are the fuel cells, which you only use for 5 or 6 hours. So when you include the capital cost penalty, the cost increases. So be careful about that.

9. *What kinds of federal research would help industry in changing processes so that they become far more efficient (for instance by requiring far less process heat), or should the federal government leave such end-use research to industry?*

There is lots of room for research priorities for industry. The federal research priorities in the U.S. energy plan suffer from a lack of focus. Understanding the fundamentals and improving the efficiency of those are good areas for government research. Maximizing carbon capture in algae is also a possibility. The key point probably is that Federal R&D is most appropriate in the areas of basic research and early development of new technology that would not otherwise be developed by private companies.

The federal research priorities should not be in efficiency of existing technologies, but on the fundamentals of the energy production industry. Efficiency ideas will come from innovation in industry.

Arjun: In your comments on the outline of my report, you were not warm to the idea of government procurement of key technologies as a way of stimulating the market.

John: Procurement – it never seems to work – it gets spent in politically correct ways or on socially wishful thinking. If there are state programs to recycle material that should not be recycled, that should be done. If it is done correctly, using the federal dollar to prime the market would be a good idea.

Arjun: What about the a commission like the military bases commission as a way of priming the market and avoiding earmarks and pork barrel type of procurement?

Dawn: Well, the base closing decisions aren't just accepted. They also get politically changed.

Arjun: I see the point of your objections more clearly. The problem of politicization of procurement seems difficult to overcome.

John: No one can disagree with [the idea of] federal leadership – but federal leadership always gets misguided due to being politicized.

10. *Does DuPont have any project that would grow biomass as part of wastewater treatment, thus helping clean the water as well as producing fuel?*

John and Dawn: Our waste disposal facilities are very small compared to municipal waste water treatment – they will do it before we do it. Also their wastewater is much richer in nutrients. You can see a living example of that – with City of Philadelphia – the discharge to the Delaware River – there are now wetlands there that have grown up around the treatment plant. It is a rich and green and wonderful nature sanctuary. Your point about using wastewater to grow biomass seems something like that. But would a municipal waste facility be better than the mouth of the Mississippi? Those are technologies that would demonstrate effectiveness in certain kinds of weather, etc. If it is not effective at municipal waste treatment plants, then there is no hope that it would be effective in industry. A city in the south should have a great advantage over any industry for trying this out. Here some combination of federal and city or state projects is a leadership that could be done. Florida would be a good place to do it.

11. *I noted in the USCAP report that there should be mechanisms for credit for those who take early action, that is before caps are imposed. I agree. The framework I am thinking of is somewhat different initially from the report, which proposes some free CO_2 (equivalent) allowances. Free allowances have created lots of problems in Europe, including issues relating to new entrants into the marketplace. I suggest: auction all CO_2 (or CO_2-equivalent) allowances for large users, including large electricity generators, for two-year periods at a time, with caps going down every two years. This will automatically benefit those who have taken early action and the new entrants with low-CO_2 footprints. For an additional benefit, I suggest that a part of the score assigned for federal and state contracting (perhaps 10%) be assigned according to the projected CO_2 emissions for the job, based on company documentation, so that all those who have a low CO_2 footprint will have a leg up. Do you have any more comment though we've covered this some already?*

John: The auction system would work in the industrial sector, small or large – it could be applied across the whole sector. But efficiency standards would work better in residential and transportation. In automobiles, I am fan of efficiency standards. I am not in favor saying John Q Public is exempt from them [standards] but business has to comply.

Arjun: Thanks so much. I'll send you these draft notes for review and correction.

[The notes were sent to Dawn Rittenhouse and John Carberry and the corrections were incorporated. This is the corrected and approved record representing the substance of the conversation.]

APPENDIX C: JAPAN FOCUS INTERVIEW ON CARBON-FREE AND NUCLEAR-FREE

Note: Mark Selden, Editor of *Japan Focus* interviewed Arjun Makhijani about this book. The interview sets the work in an international context. It is reproduced here, slightly edited, with permission. *Japan Focus* is a web-based journal, located at www.japanfocus.org.

Why zero carbon emissions? Not even the boldest proposals have called for zero emissions, even defined as you do as a few percentage points of CO_2 emissions on either side of zero. We understand the necessity to sharply reduce carbon emissions to safe limits and to reverse the carbon excess in the environment. Still, why zero emissions? Is this simply a means to draw attention to the problem where substantial reductions rather than zero emissions would solve the multiple problems associated with the present profligate fossil fuel and other nonrenewable energy consumption? Does the demand for zero emissions not risk alienating potential support for a feasible program of sharp reductions?

The United Nations Framework Convention on Climate Change requires the burden of reductions to be borne with present and past inequities taken into account. At the very least, this will mean that any CO_2 emissions that are allowed would be allocated on a per person basis.

At the same time, the Intergovernmental Panel on Climate Change has estimated that if temperature rise by mid-century is to be limited to less than 2 to 2.4 degrees Celsius, it will be necessary to reduce global CO_2 emissions by 50 to 85 percent. The former number (a 50 percent reduction in emissions) corresponds to a 15 percent chance that the temperature rise will be limited to that range; the latter (an 85 percent reduction in emissions) an 85 percent chance. If the remaining CO_2 emissions are allocated on a per person basis, and we assume that we will need a reduction of 50 percent in CO_2 emissions, the United States will have to reduce its emissions by 88 percent. At this level, it will still be very likely that we will not be able to meet the temperature rise limit. For that we must reduce global emissions by 85 percent. The U.S. goal, given its world-leading position in CO_2 emissions, would then have to be 96 percent. This is operationally the same as zero-CO_2 emissions. (I assume a global population of 9.1 billion and a U.S. population of 420 million in the year 2050).

The other reason to actually go to 100 percent elimination is that climate change is shaping up to be more severe than estimated by models. We may have to remove CO_2 from the atmosphere that has already been emitted to try to mitigate the severity. It makes no sense to remove CO_2 at great expense while emitting more. So I studied the technical feasibility of achieving an energy economy actually eliminating all fossil fuels. Some coal and natural gas infrastructure would be maintained as a contingency, but not used unless there is a major technical failure. Even then coal would only be used with carbon sequestration.

Finally, the solution to other problems, notably oil-related insecurities accompanies a zero-CO_2 economy. It is not necessary to have a zero-CO_2 economy in the United States to accomplish a reduction of oil-related insecurities. There are a variety of ways to do that, such as turning coal to liquid fuels. But such choices would aggravate CO_2 emissions.

You focus on the U.S. Could you locate the U.S. within the global framework of energy consumption, showing the critical dimensions of U.S. reduction of carbon emissions to the overall future of humanity? In particular, could you locate the U.S. problem within the framework of the Asia Pacific region?

I focus on the U.S. because it is the largest emitter of CO_2 as of 2004, the reference year for this study. But obviously it makes no sense for the U.S. to eliminate all its CO_2 emissions, while others are doing business-as-usual and continuing fossil fuel use.

A U.S. direction of significantly reducing petroleum consumption would have a major positive effect on global politics, including in the Asia Pacific region. Much geopolitical competition, including between China and Japan, is over oil. This is exemplified in their dispute over rights to oil resources in the Sea of Japan, in competing plans for the location of Russia's oil pipeline, and in territorial conflicts over the Spratly Islands involving several Asian countries. Some U.S.-Chinese tensions are also related to oil, including their competition in Africa and their differing stance toward Iran. If there is less reason for Japan and China to compete over petroleum, the drift towards a more active military posture by Japan may also be halted.

I am not saying that a gradual U.S. withdrawal from the oil market would solve most or all major geopolitical problems, but it could contribute to a different setting in which other problems are addressed. New problems may also emerge. For instance, oil exporting countries may want to be compensated for not producing oil.

Finally, a U.S. goal of zero-CO_2 emissions would bring China and India to the table of climate change discussions in more positive ways, which would benefit the whole Asian Pacific region and the world.

One notable omission from your recommendations concerns the vast global oil and energy uses of the Pentagon, by far the largest U.S. energy consumer. Please comment on the reasons for the omission, and suggest how you would approach this important element in any emissions reduction program.

The Pentagon's oil consumption is quite high. Direct Pentagon oil demand was about 320,000 barrels of oil a day in 2006.[1] But this is mainly a reflection of the Pentagon budget, which is now about $650 billion per year. This amounts to about 5 percent of the U.S. Gross Domestic Product. The U.S. consumes about 20 million barrels of oil a day; five percent of that is 1 million barrels a day. So, while 320,000 barrels a day looks large, it is a smaller proportion of oil than the Pentagon budget is of U.S. GDP. Actually, it does not include all Pentagon oil consumption because it takes no account of the oil used by Pentagon contractors and the companies that build U.S. military equipment.

The underlying problem is not really high oil consumption, though there are probably inefficiencies in the Pentagon as in most other sectors of the economy. The real issue is high military spending. Oil consumption is a reflection of that. The issue of military spending is important, but it is not within the scope of the zero-CO_2 emissions book that I have just finished.

A vigorous carbon emissions reduction program on even a fraction of the scale your report envisages would enable the U.S. to lead the international drive to overcome global warming, reversing its present position as a laggard in this arena. I understand the necessity to issue a wakeup call to the U.S. Nevertheless, what considerations led you to focus exclusively on the U.S. rather than locating the problem in interactive terms involving other nations and international organizations?

I think that without US action, there can be no US leadership, and without such leadership, global efforts to curb emissions will be gravely weakened. At this stage, preaching temperance from the barstool is not an option for the U.S., if it ever was. As I have already explained, a zero-CO_2 emissions goal is not only desirable for protecting the environment, it is also implied by U.S. treaty commitments. It will be impossible to bring China and India and Brazil and other developing countries to the table for really serious reductions in CO_2 emissions, unless the US abides by the spirit of the United Nations Framework Convention on Climate Change. And that needs to happen soon. I believe that is why former Vice-President Gore has called on the developed countries to reduce their greenhouse gas emissions by 90 percent by 2050.[2] It will be interesting to see how President Bush's climate change summit at the end of September develops, and what India and China will have to say.

There are technical imperatives if we are to save the earth, but there are also political imperatives. How can we frame a series of proposals that will be taken seriously by political actors? Recently, Australian environmentalist, Clive Hamilton, critiqued George Monbiot's call for Britain to reduce carbon emissions by 90 percent by 2030 as politically unrealizable, however praiseworthy. In the US, a nation with no serious debate about a feasible emissions reduction program, is your call merely a wakeup call drawing attention to the disasters that await us? Under what circumstances could it become a rallying cry for political forces in the US and internationally? All the more so with neoliberal thinking so powerfully in the ascendant, what would be required to contemplate the unthinkable proposal you have formulated?

My proposal should be distinguished from Monbiot's 90 percent reduction by 2030. That seems much too short a time for the immense investment and infrastructural change that will be needed for a 90 percent CO_2 reduction. I think it will take about 40 years to do the job. If there are several new technological breakthroughs in the next decade, it could possibly be done by 2040. Even then, I recognize that the political hurdles are immense. There is a huge lobby for fossil fuels; solar energy and efficiency are puny by comparison.

Even though President Bush has promised to "consider seriously decisions made by the European Union..." which imply global reductions in CO_2 of 50 to 85 percent,[3] were he confronted with a bill that required corresponding U.S. action (88 to 96 percent reductions by 2050), he would be likely to veto it.

The most leverage, politically and economically, is at the state and city level and with the corporations that stand to lose a lot through inaction. Cities are where much of the action needs to take place anyway. They can require the conversion of their taxis to hybrids and purchase plug-in hybrids. They can follow the lead of New York City in encouraging bicycling and car-free greenways[4] and promoting public transportation or London in restricting traffic to and from the core of the city.[5] They can lobby Congress for grants for renewable energy infrastructure. They can grow energy crops in their wastewater systems.

There are also corporations, for instance insurance companies like Swiss Re, and chemical companies like DuPont, that see the handwriting of climate change on the wall. They also want a piece of the action in research and the production of environmentally sound products. Some of them have accepted a goal of 60 to 80 percent reduction in U.S. greenhouse gas emissions.

California is in fact a leader in energy policy today. Governor Schwarzenegger aspires to be a global leader on climate change. In his State of the State address last January he said:

> Not only can we lead California into the future ... we can show the nation and the world how to get there. ...We are the modern equivalent of the ancient city-states of Athens and Sparta. California has the ideas of Athens and the power of Sparta.

> ...I propose that California be the first in the world to develop a low carbon fuel standard that leads us away from fossil fuels...Let us blaze the way, for the U.S. and for China and for the rest of the world.

> ...California has the muscle to bring about such change. I say use it.[6]

He will go to the United Nations in September and talk about climate change. The Secretary General of the UN has made it a top priority.[7]

There is a parallel to the phase-out of CFCs, which deplete the ozone layer. In the late 1980s and early 1990s, there were so many different local and state regulations on reducing CFC emissions that large corporations began to lobby seriously for national regulations. Something similar needs to happen with setting an ambitious goal for eliminating CO_2 emissions, and there are many signs that it is already happening. Basically, Washington will be forced to act by changes throughout the country. It is important to make it an issue in the next elections at all levels from the local to the presidential.

I did the study to show that it is technically and economically feasible to eliminate fossil fuels from the U.S. economy. That is a pre-condition for pushing to get it done. Of course, it does not guarantee that it will get done. It will take a lot of hard work and several years to build the political muscle for a zero-CO_2 emissions goal to be adopted. But I think it can be done.

The executive summary of Arjun Makhijani's forthcoming book, is available here: http://www.ieer.org/carbonfree/summary.pdf

ENDNOTES

Front Matter

1   Paley Commission 1952, v.IV, page 220
2   Energy Policy Act of 2005. See, for example, Title XIII and Title XVII
3   See, for example, Interior 2006 and Interior 2007. While the Bush administration has not tied the polar bear population decline to anthropogenic climate change, it is cited here because the warming climate has played a central role in it – a fact that is acknowledged in the Department of Interior press release and its Federal Register notice cited here.
4   Interior 2007
5   See NOAA CO_2 Trends.
6   USGCRP 2003
7   See, for example, Walter et al. 2006.
8   See NOAA CO_2 Trends. In addition there are other greenhouse gases. The Stern Review notes that "Greenhouse-gas concentrations in the atmosphere now stand at around 430 ppm CO_2 equivalent." (Stern Review 2007 page 193. For trends on greenhouse gases, see *http://cdiac. ornl.gov/trends/trends.htm* (CDIAC 2003-2006 Trends). In this book, "CO_2 concentration" refers strictly to carbon dioxide, while "CO_2-equivalent concentration" refers to the concentration of a combination of the most important greenhouse gases adjusted to an equivalent CO_2 concentration by a factor called the "global warming potential" which measures their impact in global warming relative to CO_2.
9   Stern Review 2007 pages 93-94
10 IPCC 2007 Table SPM.5, (page 23). See scenario AA.
11 Stern Review 2006 Executive Summary Figure 2 (page v)
12 IPCC 2007 Table SPM.5, (page 23)
13 Smith 2006. See also Makhijani 2004 and Smith and Makhijani 2006. Brice Smith was Senior Scientist at IEER when *Insurmountable Risks* was written. He continues in that role in the summers. He is now Assistant Professor of Physics at the State University of New York College at Cortland.
14 See for instance, the most recent report of the National Research Council, (NAS/NRC 2006)
15 Makhijani and Saleska 1999
16 Bush 2004 and Weisman 2004
17 NPT Article VIII
18 Qusti 2006
19 EIA CABS 2005 US
20 Kissinger 2007 and ISG 2006. See below.
21 Kissinger 2007
22 ISG 2006 page 28
23 ISG 2006 page 30 and EIA Petroleum Persian Gulf 2007
24 Yergin 1991, Chapter 10
25 A truly instructive history of oil, complete with colorful quotes from leaders in the first part of the twentieth century, can be found in Yergin 1991. For instance, Senator Bérenger of France, in 1918, noted, with some drama that oil is "the blood of victory ... Germany had boasted too much of its superiority in iron and coal, but it had not taken sufficient account of our superiority of oil.... As oil had been the blood of war, so it would be the blood of the peace. At this hour, at the beginning of the peace, our civilian populations, our industries, our commerce, our farmers are all calling for more oil, always more oil, for more gasoline, always more gasoline. More oil, ever more oil!" As quoted in Yergin 1991 page 183. Translated from the French in Yergin, with the exception of "More oil, ever more oil."
26 Bush 2006
27 Vedantam 2006
28 I wish to thank Julie Enszer for making the recycling analogy and raising the issue of what social dynamic might cause a similar change in personal energy use habits.
29 President Bush said: "America is on the verge of technological breakthroughs that will enable us to live our lives less dependent on oil. And these technologies will help us be better stewards of

the environment, and they will help us to confront the serious challenge of global climate change."
(Bush 2007). His remarks were noted around the world. See, for example, Baker and Mufson
2007.

30   G8 Climate Declaration 2007 page 15
31   For instance, DuPont reports having achieved a 72 percent reduction in greenhouse gas emissions
between 1990 and 2003, almost all of which were non-CO_2 greenhouse gases. (DuPont 2006b)
32   Gore 2007

Chapter 1: Setting the Stage

1   G8 Climate Declaration 2007 pages 15-16 (emphasis added)
2   Bush 2000 and Pianin and Goldstein 2001
3   EPA GHG 2006
4   In this study we include use of coal and organic materials in cement manufacture under the rubric
of "fossil fuels" for convenience.
5   Buckley 2007
6   The confidence interval that 50 to 85 percent CO_2 reductions will keep the temperature rise to 2
to 2.4 degrees Celsius above pre-industrial levels is 15 to 85 percent. See IPCC 2007 Table SPM.5
(page 23) and footnote d to the Table. This table specifies reductions in CO_2 alone, rather than
reductions in all greenhouse gases, in terms of CO_2-equivalent emissions.
7   A reference global population of 9.1 billion and a U.S. population of 420 million are used
throughout this book in the calculations for the year 2050. World population is from a 2006 United
Nations estimate (UN 2006). The U.S. population projection is from the U.S. Census Bureau project
(US Census 2004). Global CO_2 emissions in the year 2000 were about 30 billion metric tons; U.S.
emissions were 5.8 billion metric tons. U.S. CO_2 emissions data are from the Environmental
Protection Agency (EPA GHG 2007 Table 3-2 (page 3-2). Global emissions data for CO_2 from fossil
fuels are from the U.S. Energy Information Administration at http://www.eia.doe.gov/environment.
html under International Emissions Data (EIA IEA 2006 Table H.1co2). The figure for CO_2 from
fossil fuels (24 billion metric tons) has to be increased by about 6 billion metric tons to account
for non-fossil-fuel-related global CO_2 emissions, for instance, from non-renewable forest burning
(Hadley Centre 2005 page 12). Data for non-fossil fuel emissions are for the 1980s. Different
sources give somewhat different numbers. Precise estimates are not required for the calculations
regarding the target percentage of U.S. emission reductions presented here.
8   In some countries a reduction of land-clearing by burning forests could contribute significant
reductions in CO_2 emissions as well, but this does not apply to the United States.
9   Gore 2007
10   As noted above, reductions in greenhouse gas emissions for the United States translate almost
directly into reductions in CO_2 emissions from fossil fuels.
11   Gore 2007 (emphasis added)
12   UNFCCC 1992 page 1
13   UNFCCC 1992 Article 3, no.1 (page 4)
14   The author of this book served on the staff of the Energy Policy Project.
15   Many of the recommendations of the final report of the Energy Policy Project, A Time to Choose
(EPP 1974), were adopted into 1975 legislation, while Dave Freeman was a consultant to the
Senate Commerce Committee, and then by the Carter administration. See Freeman 2007,
forthcoming book.
16   Nuclear power supplies about 20 percent of the electricity in the United States. (EIA AER 2006
Table 8.4a) The generating efficiency is about one-third — that is, about two-thirds of the heat
generated in nuclear power reactors is discharged as waste heat into rivers, oceans, and the
atmosphere.
17   EIA AER 2006 Table 5.1
18   EIA AER 2006 Table 2.1d and value of production data derived from the Statistical Abstract of the
United States. (Statistical Abstract Online 2007 Table 897 and Table 767)
19   EIA AER 2006 Table 2.1a. All energy data are from the Energy Information Administration, unless
otherwise stated.
20   See, for example, EERE 2006b and California Energy Commission 2007.
21   Statistical Abstract Online 2007 Table 1081
22   Rosenfeld and McAuliffe 2006. Emphasis in the original.

23   EIA GHG 2006 Table ES2 (page 2) and EIA AER 2006 Table 1.3

24   In this study, we take a technical approach to the question of what services people want and do not inquire into the reasons for high material demands or alternative ways in which those demands could be fulfilled. For instance, if locally-grown food were a much larger part of the food system, it would likely save energy and probably provide a more secure food system. However, the kinds of policies, practices, and personal preferences that would be needed to make those changes are in themselves quite complex and would require a study of far greater scope than this one to address carefully.

25   Estimated by the author from Rosenfeld 2003 Figure 2.

26   Nadel et al. 2006 Figure 2.1 (page 6). In 1996 dollars. Nadel et al. defines "unit value" as "average manufacturer cost and profit."

27   DOE and EPA 2007

28   USGS 2006

29   Hu and Young 1994 Table 7.16 (page 7-26)

Chapter 2: Broad Energy and Economic Considerations

1   EIA IEA 2006 Table E.1p

2   EIA AEO 2006 Table 1 (page 11)

3   In an interview, Dawn Rittenhouse and John Carberry of DuPont indicated that a one to two percent absolute decline per year was a reasonable energy efficiency goal under a system that capped emission allowances seeking to reduce greenhouse gas emissions greatly from the present-day level. (Rittenhouse and Carberry 2007–See Appendix B)

4   Smith 2006 Tables 2.2 and 2.3 (pages 35 and 36), which were estimated from MIT 2003 and University of Chicago 2004. Costs of power plants are estimated as of 2003 or 2004.

5   Hong and Slatick 1994. A heat rate of 10,000 Btu/kWh is assumed. (EIA Kyoto 1998 Table 17 (page 75)

6   Fritsche and Lim 2006 Figure 4 (page 6)

7   The experience with CO_2 sequestration so far and various cost estimates for sequestration associated with IGCC plants are summarized in Smith 2006 pages 89 to 96.

8   The insurance problem was pointed out to me by Isaac Berzin, Chief Technology Officer of GreenFuel, a company formed to capture power plant exhaust CO_2 in microalgae (see subsequent chapters, including Chapter 3). (Berzin 2007)

9   Schrag 2007

10   MIT 2003 Table 5.1, 25-year- and 40-year levelized costs

11   NCUC 2007 page 213

12   Based on Makhijani 2001 page 30. The estimated added cost of the French program is $800 million for 20 reactors, each using plutonium fuel in 30 percent of the core, over and above the cost of uranium fuel. This amounts to about 2 cents per kWh added cost for the electricity generated using that fuel.

13   The heat rate for a coal-fired power plant assumed = 10,000 Btu/kWh, which represents an efficiency of about 34 percent. This is somewhat higher than the average at present, but lower than new coal-fired power plants. A detailed paper published by the Energy Information Administration indicates CO_2 emission factors between 205 and 227 pounds of CO_2 per million Btu of coal. (EIA 1995 Coal) We have used 215 pounds per million Btu in this calculation, which when rounded yields about $10 per metric ton of CO_2 per cent per kWh at the selected heat rate. The standard emission factor for electric utilities published by the EIA of about 26 million metric tons of carbon per quadrillion Btu for 2002 yields approximately the same result. See EIA factors at EIA GHG 2005 Docs Table 6-1 (page 187).

14   Berzin 2007

15   In this context we will not consider single stage natural gas turbines, since the avoided cost for combined cycle power plants and hence the imputed CO_2 cost is smaller.

16   EIA AER 2006 Table 6.8

17   EIA Electric Power 2006 Table 2.8 (page 23). This is the total of combined cycle and single stage gas turbine capacity.

18   Estimated from total capacity and EIA data for electricity generation. (EIA Electric Power 2006 Table 1.1 (page 13))

19   A heat rate of 7,000 Btu per kWh is assumed for a natural gas combined cycle plant. (EIA Kyoto 1998 Table 17 (page 75)) This corresponds to an efficiency of just under 50 percent.
20   For a discussion of wind-generated electricity costs, see Makhijani et al. 2004.
21   Light 2003. The scenario studied in Light is that of actual electricity generation in fuel cell vehicles. The main cost in that case is that of the fuel. In the example considered here, the batteries in an all-electric vehicle are used for storage and retransmission into the grid. The costs, therefore, are those of the V2G infrastructure plus the electricity losses in charging and discharging.
22   A paper on lithium-ion batteries (Buchmann 2006) and a company that makes lithium-ion batteries for solar racing cars (Solion 2003) claim 99 percent efficiency. However, Tesla Motors provides a figure of 86 percent. See Eberhard and Tarpenning 2005. For further discussion see Chapter 3.
23   For the purposes of this discussion, we ignore the potential for negative CO_2 costs. In effect, we are assuming that policies that will be in place, including CO_2 caps, will cause the adoption of technologies that are profitable even without CO_2 caps (see Chapter 6).
24   See, for instance, Paul 2002 and Escobar 2001.
25   EIA CABS 2006 Oil Prices and EIA STEO 2006
26   COS-Trust 2007, which gives a detailed financial evaluation, estimates the cost at $36.83 (Canadian dollars) per barrel or about $32 (U.S. dollars) per barrel. We will use a range of $30 to $35 (U.S. dollars) per barrel in this report.
27   ISG 2006 page 30, EIA Petroleum Persian Gulf 2007, and EIA Petroleum 2007
28   See, for example, EIA GHG 2006 page 20.
29   EIA Spot Prices at http://tonto.eia.doe.gov/dnav/pet/pet_pri_spt_s1_d.htm, viewed on July 3, 2007
30   EIA Gas Primer 2006
31   Google 2007, viewed early July 2007. The mileage varies somewhat from time to time depending on the specifics of the use of the cars.
32   DuPont 2006
33   I would like to especially thank Hisham Zerriffi, one of the project's Advisory Board members, for pointing to the necessity of developing technologies that would allow large-scale removal of CO_2 from the atmosphere to be a realistic option. Some approaches and policies are discussed in subsequent chapters of this report.

Chapter 3: Technologies—Supply, Storage, and Conversion

1   AWEA 2006
2   The idea of how to illustrate this problem comes from Walt Musial (Musial 2005 Slide 2).
3   GWEC 2007
4   Parsons et al. 2006 page 3
5   Parsons et al. 2006 page 7
6   EnerNex 2006
7   Musial 2005 Slide 10. This estimate excludes Alaska.
8   EIA AER 2006 Table 8.2a. A 35 percent capacity factor is assumed for offshore wind.
9   Keith et al. 2004
10   Makhijani et al. 2004. For an actual example of a wind farm see Kimball 2004 which has a capital cost of $1,330 per kW. A survey of costs in 2003 is available at Public Renewables.
11   Solar energy land-area data are generally provided in metric units and we retain that convention here. One square meter equals about 1.2 square yards or about 11 square feet.
12   DOE 2006
13   DOE 2006
14   Kemp 2006
15   NREL 2004. NREL's research achieved a record 16.5 percent thin film solar PV efficiency in 2001. See Wu et al. 2001.
16   Data about Nanosolar are from its web site, www.nanosolar.com. The timeline with links to more details is at http://www.nanosolar.com/history.htm. For the interview see earth2tech 2007.
17   First Solar 2007 and Fairley 2007
18   See Smith 2006 pages 83-85 for a summary of some recent developments. See also Eskom 2005.
19   Feldt 2006
20   Evergreen Solar 2006, Evergreen Solar 2006b and Evergreen Solar 2006c

21 DOE 2007a

22 Gas turbine cost of $250/kW, operating 300 hours per year. Assumptions: heat rate 13,000 Btu/ kWh, gas cost = $8/million Btu

23 Keshner and Arya 2004

24 PowerLight 2002

25 Google Blog 2006

26 Earth Policy 2001. The estimate assumes three parking spots per vehicle and 30 square meters per parking spot. The area per parking spot appears to include both the area of the spot itself as well as associated paved surface needed for movement of vehicles in parking lots.

27 Google has set up a special web site for its plug-in hybrid program, which includes the V2G test. Google 2007

28 EIA 1995 Renewables pages 101-102, 109

29 Herrmann, Geyer, and Kearney 2002 slide 21

30 Fthenakis and Kim 2006

31 Fritsche and Lim 2006 Figure 4 (page 6)

32 Fthenakis and Kim 2006 Table 1 (page 3)

33 Smith 2006 has raised these points in his discussion of solar PV. See pages 84-85.

34 Makhijani and Poole 1975 and Makhijani 1990. Biomass used as food for draft animals is one of the largest energy inputs in parts of rural South Asia, for instance. Yet it is not included in compilations of energy data. It will be important to do so in assessing issues of energy, food, land use, and social and economic justice as the climate debate becomes more intense in regard to Asian developing countries.

35 Sandalow 2006 page 67

36 Bush 2006 and Bush 2007

37 Ethanol Market 2007. Note that ethanol has smaller energy content than gasoline. A gallon of ethanol is equal in fuel value to about 0.61 gallons of gasoline.

38 Bush 2007

39 EERE Solar 2007

40 See NREL map above, Figure 3-6.

41 Typical yields for corn are used. See Farrell et al. 2006b Figures S1 and S2 (pages 14-15). A switchgrass yield of 13,000 kilograms per hectare is assumed. See Farrell et al. 2006 and Farrell et al. 2006b. A value of 5 kWh per square meter day is used for typical incident solar energy. See Figure 3-6.

42 This assumes an energy value of corn of 18 million Btu per metric ton, incident solar energy of 5 kWh per square meter per day, and one crop per year.

43 Farrell et al. 2006b page 4 and Table S3 (page 21)

44 Many studies yielding different results have been done. Farrell et al. 2006 does a careful analysis of six studies and compares the methods and results. Farrell et al. 2006 and the supporting material in Farrell et al. 2006b are used here to provide the basis for the results shown. All figures are rounded and approximate, since that suffices for the purpose of illuminating broad policy directions and concepts for a zero-CO_2 economy and its implications for present policy direction. Gasoline emissions, like ethanol emissions, were computed on a lifecycle basis in Farrell et al. 2006. Overall, a small reduction in greenhouse gas emissions appears to result from corn-derived ethanol, when the energy and emissions credits for the co-products are taken into account.

45 Malkin 2007

46 Runge and Senauer 2007

47 Runge and Senauer 2007

48 Buckland 2005 and Rosenthal 2007

49 Delft Hydraulics 2006 page 30

50 We do not address issues related to the Brazilian ethanol from sugarcane here. It has generally been considered that this has a positive effect on greenhouse gas emissions. However, this does not take into account the overall changes in land use patterns of which ethanol production is a part. The total, direct and indirect, effect of food crops for export, ethanol for fuel, and providing for a growing population with higher incomes creates pressures on the land whose net effect, for instance, on deforestation in the Amazon region is difficult to determine even though sugarcane is not cultivated on cleared Amazon forest land. Further, fuel crops could be grown on cleared forest land. As Farrell et al. have pointed out in the context of potential imports of ethanol into the United States: "The possibility of importing ethanol suggests that land use changes as a result

of U.S. ethanol use could occur outside of the country, raising concerns about, for instance, the conversion of rainforest into plantations for fuel production. Estimating the magnitude of such effects would be very difficult, requiring analysis of land productivity and availability, commodity markets, and other factors...." (Farrell et al. 2006b page 12). Importing large amounts of ethanol or other fuels made from food crops or importing foodstuffs into the West from developing countries for the purpose of producing fuel is likely to have a deleterious effect on poor peasants and landless laborers and other people living in poverty or close to poverty in developing countries. See Runge and Senauer 2007.

51 The total energy content of all crop residues in the United States is about three quadrillion Btu, or less than 5 percent of the natural gas and petroleum use. (Milbrandt 2005 Figure 28 (page 47)). Only a fraction of this would be available for fuel production if appropriate attention is paid to soil conservation issues.

52 Berzin 2007 estimates a productivity of 100 grams per square meter per day for very sunny areas like Arizona. This translates into 250 metric tons per year on the basis of 250 sunny days per year. The productivity depends of the type of microalgae and the circumstances in which they are cultured. See NREL 1998.

53 MIT News 2004 and MIT Cogen 2007

54 Berzin 2007

55 CK Environmental 2004

56 Berzin 2007 and Bane 2007

57 Berzin 2007

58 Berzin 2007

59 Berzin 2007

60 We will use a value of 18 million Btu per metric ton of dry biomass, also called "Bone Dry" biomass, throughout this report. While the figures vary somewhat from one form of biomass to another, the use of a single value is justified given the approximate nature of the calculations. Inferred from values for switchgrass (NREL 2005 Figure 28 and Table 5).

61 Berzin 2007

62 Greene et al. 2004. See page 63 for a discussion of output.

63 Greene et al. 2004 page vi

64 Greene et al. 2004 Table 5 (page 26) and discussion on pages 25 and 26

65 Farrell et al. 2006 and Farrell et al. 2006b

66 Farrell et al. 2006b Table S3 (page 21). One megajoule is about 950 Btu. One gallon of gasoline is about 125,000 Btu.

67 Tilman, Hill, and Lehman 2006

68 Wolverton and McDonald 1977

69 Wolverton and McDonald 1979 page [2]

70 EPA 1988 page 48

71 The rest of this account of the NASA project in Bay St. Louis is based on Wolverton and McDonald 1977.

72 Wolverton and McDonald 1977 page 207

73 EPA 1988. The rest of the discussion is based on this EPA overview and design document, unless otherwise specified.

74 See, for instance, Wolverton and McDonald 1979.

75 See, for instance, Moreland and Collins 1990.

76 DOE 2007

77 DOE 2007 Table 3.1.10

78 DOE 2007 Table 3.1.9

79 EPRI 2005

80 This section is based on MIT 2006.

81 First called the Solar Energy Research Institute.

82 MIT 2006 page 1-6

83 EIA 1995 Renewables page 109

84 In this study, we are not considering new pumped hydropower storage, which uses off-peak power from a source other than hydroelectric power plants to pump water downstream of a dam back into the reservoir. The water is then used to generate electricity at times of peak demand. The capacity for new storage would likely be limited in the context of very large-scale

85 implementation of solar and wind energy.
This would apply to fleet vehicles with charging equipment that can carry large currents. Phoenix Motorcars, Inc. is manufacturing SUV pickup trucks for such applications with a 10-minute charging time. See http://www.phoenixmotorcars.com/.

86 Based on a source in industry and Miller 2007. Lithium-ion battery costs vary and are more than $500 per kWh. Installed costs of battery systems in cars can be well above $2,000 per kWh due to very small-scale (one to a few cars) custom installation.

87 Kempton and Letendre 1997

88 University of Delaware V2G

89 We have focused here on batteries since plug-in hybrids and lithium-ion all-electric vehicles are much closer to commercialization than fuel cell vehicles.

90 See "Recharge a Car, Recharge the Grid, Recharge the Planet" at Google 2007. For the lithium-ion battery type being used by Google, see Hybridcars.com 2007.

91 Light 2003

92 Eberhard and Tarpenning 2006 page 2 and Solion 2003

93 A fact sheet on the battery is available on the company's web site at www.altairnano.com/documents/NanoSafeBackgrounder060920.pdf. (Altairnano 2006). See O'Shea 2006 for a trade journal news report on the final performance test.

94 Not all hybrid cars have the capacity to run on electricity only. The most common one, the Toyota Prius, does.

95 See http://www.calcars.org/carmakers.html#vvquotes at the web site of Calcars, a non-profit that promotes plug-in hybrids.

96 Miller 2007

97 Miller 2007. AFS Trinity Power aims for a liquid fuel efficiency of 150 to 250 miles per gallon (plus electricity enough to drive 40 miles on the battery alone) (AFS Trinity 2006)

98 Experimental work on these capacitors is currently being carried out at MIT, among other places. See Schindall 2007 and MITLEES 2006

99 Shepard and van der Linden 2001 and CAES McIntosh. These are the sources for the following paragraphs.

100 McIntosh Project web site at http://www.caes.net/mcintosh.html. (CAES McIntosh) A heat rate of 10,000 Btu/kWh for coal-fired power plants is assumed.

101 See Energy Services 2003.

102 The energy sector emitted about 6 billion metric tons of CO_2 per year in 2005; the other greenhouse gases account for about 1 billion metric tons per year CO_2 equivalent.

103 Wilson, Johnson, and Keith 2003 page 3476

104 Utah Geological Survey

105 Utah Geological Survey

106 Berzin 2007

Chapter 4: Technologies—Demand-Side Sectors

1 Data for these and other efficient buildings are at http://www.eere.energy.gov/buildings/database/index.cfm. (EERE 2004). This web site provides links to a wealth of material describing energy efficient equipment and design concepts and a glossary at http://www.eere.energy.gov/consumer/information_resources/index.cfm/mytopic=60001.

2 Quote from EERE Hanover 2002. Many design features are described on the Web at http://www.eere.energy.gov/buildings/database/energy.cfm?ProjectID=49.

3 The 58,000 Btu per square foot is calculated from EERE 2006 Table 1.2.3 and EIA AEO Assumptions 2006 page 23 and EERE 2006 Table 2.1.1

4 Winkler, 2007

5 Quote from EERE Takoma 2003 Energy. Many design features are described on the Web at http://www.eere.energy.gov/buildings/database/overview.cfm?projectid=70.

6 Quote from EERE Durant 2007. Many design features are described on the Web at http://www.eere.energy.gov/buildings/database/energy.cfm?ProjectID=46.

7 EERE Cambria 2002

8 Sachs et al. 2004 page 40. "Standby power is the electricity consumed by end-use electrical equipment that is switched off or not performing its main function." (Sachs et al. 2004 page 40)

9 The details of this project are from Parker, Sherwin, and Floyd 1998, unless otherwise mentioned.

10   Sunlight Direct 2005
11   See the web site of the Oak Ridge Solar Technologies Program at
      http://www.ornl.gov/sci/solar/. (ORNL Solar 2007)
12   Narendran et al. 2005
13   Tesla Motors 2007 and Phoenix Motorcars. Tesla motors uses commercial lithium-ion batteries
      in a large battery back specially developed for automobiles. Phoenix Motorcars uses new
      nanotechnology lithium-ion batteries.
14   A European Union survey of hydrogen fuel for aircraft can be found in links to documents at
      European Commission 2000.
15   DARPA 2006
16   Tupolev 2006
17   It should be noted that most people in the infamous Hindenburg disaster survived. There is still a
      considerable controversy over the causes of the accident and fire, with an excellent survey found
      at Wikipedia Hindenburg 2007.
18   European Commission 2000. In an interesting research project, Georgia Tech has done test flights
      of an unmanned 500 watt hydrogen fuel cell powered plane for one minute at a time. (Georgia
      Tech 2006)
19   Airbus Deutschland 2003 page 5
20   Airbus Deutschland 2003 page 12
21   Airbus Deutschland 2003 pages 29-30
22   Airbus Deutschland 2003 page 65
23   Airbus Deutschland 2003 page 47
24   The fraction is difficult to read from the bar chart, but appears to be about 5 percent.
25   Airbus 2001 Slide 11
26   O'Neill 2006
27   Bloomberg 2007 Figure i
28   Environmental Defense 1999 shows a timeline of environmental justice struggles in Los Angeles,
      which includes the public transit bus story.
29   Rundle et al. 2007
30   See Appendix B (Rittenhouse and Carberry 2007).

Chapter 5: A Reference Zero-CO$_2$ Scenario

1   EERE 2006 Table 2.1.1 (page 2-1) for 2004. The number for 2050 is calculated.
2   The main efficiency and technology assumptions for the year 2050 for the residential sector are: 1.
     Overall building envelope heating requirement reduction relative to business-as-usual: 40%.
     2. Heating technologies: conventional, similar to natural gas forced air or circulating hot water
     and geothermal heat pumps, one-third each; solar thermal assisted fuel or electricity,
     solar portion of the load 13%; CHP (combined heat and power, mainly apartment buildings), 20%.
     3. Cooling system efficiency: among the higher efficiency systems available today (coefficient
     of performance = 6, or SEER = about 20). 4. Hot water: solar thermal portion of the load = 40%.
     The same end result can be achieved with different combinations of HVAC and water heating
     technologies. Other appliance efficiency, factor of 2 improvement over that projected in the
     business-as-usual scenario. Note that the effect of standards for refrigerators, for instance, in
     thirty years has been an improvement by a factor of 3 to 4. These assumptions are based on a
     survey of the literature of efficient buildings and residential sector technologies.
3   The main efficiency and technology assumptions for the year 2050 for the commercial sector are:
     1. Overall building envelope heating requirement reduction relative to business-as-usual: 30%.
     2. Heating technologies: geothermal heat pumps: one-third each; solar thermal assisted fuel
     or electricity, solar portion of the load 15%; CHP (combined heat and power), 25%. 3. Cooling
     system efficiency: coefficient of performance = 6, or SEER = about 20, plus use of absorption
     air-conditioning for 25 percent of the load. Building envelope and lighting improvements reduce
     cooling load by 30% relative to business-as-usual. 4. Hot water: solar thermal portion of the
     load = 40%. Balance electricity and fuel, including that associated with CHP systems. 5. Lighting
     and other appliance electricity use a factor of 3 lower than business-as-usual – largely due to
     efficiency improvements in lighting. These assumptions are based on a survey of the literature of
     efficient buildings and commercial sector technologies, such as LED lights of new designs and
     solar-hybrid lighting.

4   Based on performance data on the web sites of Tesla Motors (www.teslamotors.com) and of Phoenix Motorcars (http://www.phoenixmotorcars.com), and an industry interview (anonymous).
5   NJC 2007
6   See http://www.teslamotors.com/media/press_room.php?id=29 and http://www.teslamotors.com/media/press_room.php?id=573, viewed on August 1, 2007.
7   Gates 2007
8   1. Light duty vehicle (less than 8,500 pounds) efficiency for new liquid fuelled vehicles: 75 miles per gallon; for new electric vehicles, 11 miles per kWh. 2. Commercial light truck efficiency is assumed to improve relative to 2004 proportionally the same as for the light duty vehicles. 3. Freight trucks, liquid fuelled: 10.7 miles per gallon; electrical (including as part of plug-in system): 1.7 miles per kWh. 4. Aircraft efficiency = 150 seat miles per gallon.
9   See, for instance, Greene et al. 2004.
10   http://www.us-cap.org/
11   The most important index of reliability of an electricity system is its "loss of load probability" or LOLP. Optimization refers in part to minimizing costs for a given level of reliability.
12   We assume only 10 kW per vehicle, even though the total available power would be considerably larger. This is because a moderate power supply level would allow the vehicle to supply energy for a longer time.
13   See the webpage of Ice Energy at http://ice-energy.com/. Example installations are cited at this web site
14   Zwetzig 2007
15   Winkler 2006
16   Zagórze, no date
17   Based on NYSERDA 2005
18   Statistical Abstract Online 2007 Table 828

Chapter 6: Options for the Roadmap to Zero-CO_2 Emissions

1   A kilogram of hydrogen is approximately equivalent in energy terms to a gallon of gasoline.
2   This section is based on DOE 2007 unless otherwise mentioned. See especially Tables 3.1.4 and 3.1.4A and the notes to these tables.
3   "This figure was created and prepared by an employee of the Midwest Research Institute (MRI) as work sponsored by an agency of the United States Government. Neither MRI or the United States Government nor any of their employees make any warranty, express or implied, or assumes any legal liability or responsibility for the accuracy, completeness, or usefulness of any information, apparatus, product, or process disclosed, or presents that its use would not infringe upon privately owned rights. The reference herein to any specific commercial product, process, or service by trade name, trademark, manufacturer, or otherwise, does not necessarily constitute or imply its endorsement, recommendation or favoring by the United States Government or MRI." –Source: National Renewable Energy Laboratory
4   Hydrogen Cars Now 2006 Gain 2006
5   See footnote L to Table 3.1.4A in DOE 2007.
6   Ford 2004
7   The land-area requirements of wind energy are very small compared to biomass cultivation for liquid biofuels. See Chapter 5.
8   Solar cooling uses an absorption air conditioning system. This is similar to systems that use waste heat for producing a cooling effect, except that the source of heat in this case is a solar energy. Pumps are used for circulation of cool water.
9   ClimateMaster Model Tranquility 27.
10   It may also be possible to use other approaches, notably flywheels. At present the use of flywheels is indicated for short-term storage needs rather than the application under consideration here – which is storage of several hours' worth of electricity supply.
11   We assume 5 kWh per day of generation per peak kW, $200 per kWh storage cost and $200 ancillary equipment capital costs. This would be typical of sunny areas. The same storage capacity would suffice for more than one day's generation in less favorable areas.
12   Siemers 2007 described the plant proposed to be built in New Mexico and also cites a skeptic. The technology has not been used on a commercial scale as yet to produce raw material for new tires.
13   Ironically, France imports all of its uranium. Its energy 'independence" in terms of proportion

of energy supply imported has actually declined – only 15 percent of the energy supply was domestically produced in 2000 compared to 22 percent in 1973. However, France's energy security in the sense of diversity and security of energy supplies has increased. But nuclear power has brought its own vulnerabilities. (Makhijani and Makhijani 2006 pages 34-37)

[14] Stern Review 2006 Executive Summary page i

[15] Throughout this analysis, we assume that policies in the direction of greater efficiency will be in place. See Chapter 7.

[16] Personal vehicles accounted for about 19 percent of total CO_2 emissions in the year 2000 and electric utilities were responsible for another 37 percent (EPRI 2005b). Residential and commercial electricity accounts for just over 70 percent of total electricity consumption. Based on these data, about 45 percent of total CO_2 emissions come from residential and commercial electricity use and personal automobiles (including SUVs and light trucks).

[17] Winkler, 2006

[18] This section is based on Makhijani and Gurney 1995, unless otherwise noted.

[19] The text of the Vienna Convention can be found at http://ozone.unep.org/ pdfs/ viennaconvention2002.pdf; viewed on 3 August 2007.

[20] Makhijani, Makhijani, and Bickel 1988

[21] See Makhijani and Gurney 1995, especially Chapters 12 and 13.

[22] Landfill gas (methane is one of the gases created by decay of the organic materials dumped in landfills) and other waste materials could also be used as energy sources. However, waste, including household and commercial municipal waste, can only meet a small fraction of energy requirements and therefore is not dealt with in the context of this report. Yet, the recovery and use of landfill gas is particularly important for global warming since it captures a greenhouse gas and provides a substitute for a fossil fuel.

Chapter 7: Policy Considerations

[1] See EPA Fact Sheet at SO_2.

[2] A comparative description along with the results can be found in Oliver 2006.

[3] Anderson 1999 and the Acid Rain Program SO_2 Allowances Fact Sheet on the web at http://epa. gov/airmarkets/trading/factsheet.html#what (EPA Fact Sheet SO_2).

[4] CCAP 1999 page 21

[5] Öko-Institut 2005 page 12

[6] CCAP 1999 page 21

[7] EPRI 2005b

[8] USCAP 2007 page 5

[9] See, for instance, Rittenhouse and Carberry 2007.

[10] Stavins 2005

[11] This corresponds to an increase in the cost of coal-generated electricity of about one cent per kWh and about half that for natural gas.

[12] See WGA 2006 pages 1, 36, 40, and 44

[13] WGA 2006. See also DSIRE 2007 for state by state listings of current incentives.

[14] WGA 2006 pages 40 and 44

[15] See Karppi 2002, for an example of rebates for earth-source heat pumps provided by a utility to a builder of a hotel in Long Island. Also see LIPA 2006.

[16] We have not dealt with the broader problem of CO_2 emissions associated with imported goods in this book. It is highly unlikely that the United States or any other country would go all the way to a zero-CO_2 emissions economy without a more general agreement to reduce global CO_2 emissions by 50 to 85 percent. In that context, the problem of the CO_2 footprint of imported goods may not be a significant issue.

[17] Andrews and Wald 2007

[18] As quoted in the Atlanta Journal-Constitution editorial published on August 2, 2007

[19] See for instance IEER 1999. This article contains a series of graphs prepared by the Department of Energy for the Nuclear Waste Technical Review Board. They show that the geology of the Yucca Mountain site is practically worthless in containing radionuclides, should they leak out of the containers.

[20] Safety reasons may cause earlier closures of some plants, but we have not taken that into account in this analysis.

21 USCAP 2007 page 7
22 Mufson and Cho 2007

Chapter 8: Roadmap for a Zero-CO$_2$ Economy

1   Lithium-ion batteries can be and are recycled. See Buchmann 2003..
2   Winkler 2006
3   Renewable Energy Access 2007
4   Calculated from EIA IEO 2006 Table A3
5   The Energy Information Administration projects crude oil prices to be in the range of about $36 to $100 per barrel in the year 2030. See EIA IEO 2007 Figure 17.
6   Miller 2007
7   Winkler 2006
8   The electricity costs are from http://www.eia.doe.gov/cneaf/electricity/epm/table5_6_a.html (EIA EPM 2007-08). Delivered fuel costs are based on a wellhead price of $7 to 8 per million Btu.
9   Northbridge 2003

Chapter 9: Summary

1   Based on a global population of 9.1 billion and a U.S. population of 420 million in 2050.
2   Offsets allow a purchaser to continue emitting CO$_2$ while paying for reductions in CO$_2$ by the party from whom the offsets are purchased. These may or may not result in actual CO$_2$ reductions. Even when they do, the emissions may be immediate while reductions may be long-term. Verification is difficult and expensive.
3   Qusti 2006

Appendix A: Nuclear Power

1   Section A is based mainly on the Foreword that the author wrote for Smith 2006. Section C is mainly based on a portion of Makhijani and Barczak 2007. For more details on the history of nuclear power see Makhijani and Saleska 1999.
2   Murray 1953
3   Cole 1953
4   AEC 1948 page 46
5   Bacher 1949 p. 6 and LANL Biography
6   Suits 1951
7   Makhijani and Saleska 1999 pages 67-68
8   See Makhijani 2001 for details relating to costs associated with efforts to commercialize plutonium fuel use. The uranium and plutonium can be separated with relative ease, yielding plutonium that could be used to make nuclear weapons.
9   J. Robert Oppenheimer, "International Control of Atomic Energy," in Morton Grodzins and Eugene Rabinowitch, eds., The Atomic Age: Scientists in National and World Affairs, (New York: Basic Books, 1963), p. 55, as quoted in Makhijani 1997.
10 The EPA standard is at 40 CFR 191. For the Science Advisory Board Report on carbon-14 see EPA 1993.
11 The DOE graphs are reprinted in IEER 1999. See also the quotes from DOE's peer review panel regarding corrosion in this article. For additional analysis of the corrosion issue, see Craig 2004. For the NRC's total system performance assessment standards, see 10 CFR 63.
12 Makhijani, Gunter, and Makhijani 2002
13 More complex methods of "recycling" have been proposed. For a critique of these, see Zerriffi and Annie Makhijani 2000.
14 Warrick 1999
15 PACE-University of Utah 2000
16 The Paducah plant did not make highly enriched uranium for the U.S. military program. However, some of the low enriched uranium that it made was subsequently enriched to weapon-grade levels at the DOE enrichment plant in Portsmouth, Ohio.
17 See Makhijani and Smith 2004.

18   The official description may be found at http://www.gnep.energy.gov/.
19   Article VI of the NPT requires negotiations in "good faith towards complete nuclear disarmament."
      A 1996 World Court advisory opinion stated that the NPT requires the actual achievement of
      complete elimination of nuclear weapons. See Deller, Makhijani, and Burroughs 2003.
20   This is my personal assessment. Herbert York, the first Director of Lawrence Livermore National
      Laboratory concurred with it in an interview he did with me in 2001. York 2001
21   Smith 2006, pages 38-42
22   EIA 1986 p. xv (emphasis added)
23   EIA 1986 page xvi
24   Georgia IRP 2007 pages 1-15
25   For a discussion of claims about the safety of new reactor designs and modified existing reactor
      designs see Makhijani and Saleska 1999.
26   Kennedy et al. 2006
27   NCUC 2007 page 213
28   Andrews and Wald 2007
29   As quoted in Andrews and Wald 2007
30   Smith 2006
31   Smith 2006, Section 4.4
32   Nature editorial 2007
33   Associated Press 2007
34   Godoy 2006
35   France 2003

Appendix C: Japan Focus Interview on Carbon-Free and Nuclear-Free

1   Karbuz 2007
2   Gore 2007
3   G8 Climate Declaration 2007. The declaration states that the United States will "consider seriously
     the decisions made by the European Union, Canada and Japan which include at least a halving of
     global emissions by 2050." (paragraph 49) In fact the EU goal is to limit the temperature rise to
     2 to 2.4 degrees Celsius. This implies a 50 to 85 percent reduction in CO_2 emissions. See IPCC
     2007 and European Parliament 2007, p. 1.
4   See New York City Department of City Planning at http://www.nyc.gov/html/dcp/html/
     transportation/td_projectbicycle.shtml (NYC 2007).
5   Changing modes of transport are not included in the reference scenario. However, certain changes
     help in reducing energy use and pollution. See Chapters 4 and 6.
6   Schwarzenegger 2007
7   Chea 2007

REFERENCES

10 CFR 63   *Code of Federal Regulations. Title 10—Energy. Chapter I--Nuclear Regulatory Commis-sion. Part 63--Disposal Of High-Level Radioactive Wastes In A Geologic Repository At Yucca Mountain, Nevada.* 1-1-06 Edition. Washington, DC: Office of the Federal Register, National Archives and Records Administration; United States Government Printing Office, 2006. On the Web at http://www.access.gpo.gov/nara/cfr/waisidx_06/10cfr63_06.html.

40 CFR 191   *Code of Federal Regulations. Title 40--Protection of Environment. Chapter I--Environ-mental Protection Agency. Part 191—Environmental Radiation Protection Standards For Management And Disposal Of Spent Nuclear Fuel, High-Level And Transuranic Radioactive Wastes* 1-06 Edition. Washington, DC: Office of the Federal Register, National Archives and Records Administration; United States Government Printing Office, 2006. On the Web at http://www.access.gpo.gov/nara/cfr/waisidx_06/40cfr191_06.html.

AEC 1948   United States. Atomic Energy Commission. *Report to the Congress, No. 4.* Washington, DC: AEC, 1948.

AFS Trinity AFS Trinity Power Corporation. *xh™ plug-in: The Extreme Hybrid™.* [Medina, WA]: AFST,
2006   [2006]. On the Web at http://www.afstrinity.com/extr.html. Viewed March 1, 2007.

Airbus 2001 Reinhard Faass. *Cryoplane: Flugzeuge mit Wasserstoffantrieb.* Hamburg, 06. Dezember 2001. On the Web at http://www.haw-hamburg.de/pers/Scholz/dglr/hh/ text_2001_12_06_Cryoplane.pdf. Presentation. At head of title: Airbus and H-Cryoplane.

Airbus   Airbus Deutschland. *Liquid Hydrogen Fuelled Aircraft: System Analysis.* [Hamburg: Airbus
Deutschland Deutschland], 24. September 2003. On the Web at http://www.aero-net.org/
2003   pdf-docs/20040202-final-technical-report-4-pv.pdf. Running title: Final Technical Report: CRYOPLANE: System Analysis (Publishable Version).

Airhart 2006 Marc Airhart. *Scientists Deepen Confidence in Technique to Reduce Greenhouse Gas Emissions.* News Releases and Features. Austin: Jackson School of Geosciences, Univer-sity of Texas at Austin, Nov. 15, 2006. On the Web at http://www.jsg.utexas.edu/news/ feats/2006/frio.html.

Altairnano   Altairnano. *NanoSafe™ Battery Technology.* Reno, NV: Altairnano, [September 20, 2006].
2006   On the Web at www.altairnano.com/documents/NanoSafeBackgrounder060920.pdf.

Anderson   Robert C. Anderson, with the assistance of Alan Carlin. *Economic Savings From Using*
1999   *Economic Incentives for Environmental Pollution Control.* [Washington, DC: Environmental Law Institute], June 1999. Section 3.2.3. Acid Rain Allowance Trading. On the Web at http://yosemite.epa.gov/ee/epa/incsave.nsf/437c451fb25915d5852564db00579f01/ e8448f37d3eeb89b85256636004f926e!OpenDocument. Viewed August 23, 2007.

Andrews and Edmund L. Andrews and Matthew L. Wald. "Energy Bill Aids Expansion Plans of Atomic
Wald 2007   Power." *New York Times,* July 31, 2007.

Associated   Associated Press. "TVA Reactor Shut Down; Cooling Water From River Too Hot." *Houston*
Press 2007   *Chronicle,* August 16, 2007. On the Web at http://www.chron.com/disp/story.mpl/ business/energy/5061439.html.

Atlanta Journal-Constitution 2007	Lyle V. Harris, for the editorial board. "Nuclear Power Lesson: A Free-flowing, Multibillion-dollar Era Too Risky; Three Mile Island Meltdown Can't be Forgotten." *Atlanta Journal-Constitution*, August 8, 2007. On the Web at http://www.ajc.com/opinion/content/opinion/stories/2007/08/01/nuked_0802.html.
AWEA 2006	American Wind Energy Association. *Congress Extends Wind Energy Production Tax Credit for an Additional Year.* News Room. Washington, DC: AWEA, December 11, 2006. On the Web at http://www.awea.org/newsroom/releases/Congress_extends_PTC_121106.html.
AWEA 2006b	American Wind Energy Association. *Wind Energy: An Untapped Resource.* Wind Energy Fact Sheet. Washington, DC: AWEA, [2006]. On the Web at http://www.awea.org/pubs/factsheets/Wind_Energy_An_Untapped_Resource.pdf.
AWEA 2007	American Wind Energy Association. *Wind Power Capacity in U.S. Increased 27% In 2006 and Is Expected to Grow an Additional 26% in 2007.* Washington, DC: AWEA, January 23, 2007. On the Web at http://www.awea.org/newsroom/releases/Wind_Power_Capacity_012307.html.
Bacher 1949	Robert Bacher. "Research and Development of Atomic Energy." *Science*, v.109, no.2819 (1949), pages 2-7.
Baker and Mufson 2007	Peter Baker and Steve Mufson. "Bush's Climate Change Remarks Weighed for Policy Shift," *Washington Post*, January 27, 2007, p. A01.
Bane 2007	Marc Bane (of GreenFuel Technologies Corporation). *RE: [Fwd: RE: Permissions request] - Engineering Scale in Louisiana.* E-mail to Lois Chalmers, dated July 26, 2007.
Bemis and Allen 2005	Gerry Bemis and Jennifer Allen. *Inventory of California Greenhouse Gas Emissions And Sinks: 1990 to 2002 Update: Prepared in Support of the 2005 Integrated Energy Policy Report.* Staff Paper. Publication CEC-600-2005-025. [Sacramento]: California Energy Commission, June 2005. On the Web at http://www.energy.ca.gov/2005publications/CEC-600-2005-025/CEC-600-2005-025.PDF.
Berzin 2007	Isaac Berzin. Telephone interview with Arjun Makhijani, February 13, 2007. Notes of Arjun Makhijani.
Bloomberg 2007	Jonathan Dickinson, ed. *Inventory of New York City Greenhouse Gas Emissions.* New York: Mayor's Office of Operations, Office of Long-term Planning and Sustainability, April 2007. On the Web at http://www.nyc.gov/html/om/pdf/ccp_report041007.pdf. Michael R. Bloomberg, Mayor.
Buchmann 2003	Isidor Buchmann. "Recycling batteries (BU21)." *BatteryUniversity.com*, created May 2003, last edited July 2003. On the Web at http://www.batteryuniversity.com/partone-20.htm.
Buchmann 2006	Isidor Buchmann. "Charging lithium-ion batteries (BU12)." *BatteryUniversity.com*, created April 2003, last edited March 2006. On the Web at http://www.batteryuniversity.com/partone-12.htm.
Buckland 2005	Helen Buckland. *The Oil for Ape Scandal: How Palm Oil is Threatening Orang-utan Survival.* Research Report. London: Friends of the Earth Trust, September 2005. On the Web at http://www.foe.co.uk/resource/reports/oil_for_ape_full.pdf.
Buckley 2007	Chris Buckley. China Says Exports Fuel Greenhouse Gas Emissions. [Sydney]: Planet Ark, 22/6/2007. On the Web at http://www.planetark.com/avantgo/dailynewsstory.cfm?newsid=42746. Copyright Reuters News Service.

Bush 2000   George W. Bush. "A Comprehensive National Energy Policy Delivered on September
                  29th, 2000 in Saginaw Michigan." In Chapter 2: Economic Issues of *In Their Own Words:*
                  Sourcebook for the 2000 Presidential Election, published by the Political Communication
                  Lab at Stanford University. On the Web at http://pcl.stanford.edu/campaigns/
                  campaign2000/sourcebook/sbpdf/economic.pdf (pages 62-70 of pdf).

Bush 2004   George W. Bush. *President Announces New Measures to Counter the Threat of WMD:*
                  Remarks by the President on Weapons of Mass Destruction Proliferation, Fort Lesley J.
                  McNair - National Defense University, Washington, D.C. [Washington, DC]: Office of the
                  Press Secretary, February 11, 2004. On the Web at http://www.whitehouse.gov/news/
                  releases/2004/02/print/20040211-4.html.

Bush 2006   George W. Bush. *President Bush Delivers State of the Union Address, United States Capi-*
                  tol, Washington, D.C. [Washington, DC]: Office of the Press Secretary, January 31, 2006.
                  On the Web at http://www.whitehouse.gov/news/releases/2006/01/20060131-10.html.

Bush 2007   George W. Bush. *President Bush Delivers State of the Union Address, United States Capi-*
                  tol, Washington, D.C. [Washington, DC]: Office of the Press Secretary, January 23, 2007.
                  On the Web at http://www.whitehouse.gov/news/releases/2007/01/20070123-2.html.

CAES   CAES Development Company. *Technology - McIntosh Project.* [Houston]: CAES, [no date].
McIntosh   On the Web at http://www.caes.net/mcintosh.html. Viewed March 26, 2007.

CalCars 2007   California Cars Initiative. *How Carmakers are Responding to the Plug-In Hybrid Opportu-*
                  nity. [Palo Alto, CA]: CalCars, 2007. On the Web at http://www.calcars.org/carmakers.
                  html#vvquotes. Viewed July 16, 2007.

California   California Energy Commission. Energy Efficiency Division. Buildings and Appliances Office.
Energy   *Appliance Efficiency & Appliance Regulations.* [Sacramento, CA]: The Commission, up-
Commission   dated 10 Jan 2007. On the Web at http://www.energy.ca.gov/efficiency/appliances/index.
2007   html.

CCAP 1999   Tim Hargrave. *Identifying the Proper Incidence of Regulation in a European Union*
                  Greenhouse Gas Emissions Allowance Trading System. Washington, DC: Center for Clean
                  Air Policy, [1999]. On the Web at http://www.ccap.org/pdf/4Incidence.pdf. Running title:
                  Determining the Point of Regulation in an EU Trading System, Center for Clean Air Policy,
                  June, 1999 DRAFT.

CDIAC 2003-   Carbon Dioxide Information Analysis Center. *Trends Online: A Compendium of Data on*
2006 Trends   *Global Change.* Oak Ridge, TN: CDIAC, Oak Ridge National Laboratory, U.S. Department of
                  Energy, 2003-2006. On the Web at http://cdiac.ornl.gov/trends/trends.htm.

Chang,   Audrey B. Chang, Arthur H. Rosenfeld, Patrick K. McAuliffe. "Energy Efficiency In California
Rosenfeld,   and the United States: Reducing Energy Costs and Greenhouse Gas Emissions." [February
and McAu-   21, 2007]. On the Web at www.energy.ca.gov/2007publications/CEC-999-2007-007/CEC-
liffe 2007   999-2007-007.PDF. Citing version viewed 7-10-07. Revised after February 21. At head of
                  title: Chapter will appear in Schneider, Rosencranz & Mastrandrea (eds.), *Climate Change*
                  Science and Policy (forthcoming in 2007)

Chea 2007   Terence Chea (Associated Press). "UN Chief Ban Ki-Moon Visits California." *Washington*
                  Post, July 27, 2007. On the Web at http://www.washingtonpost.com/wp-dyn/content/
                  article/2007/07/27/AR2007072700241.html.

CK   David L. Mackintosh (CK Environmental, Inc.). *Re: GreenFuel at MIT Cogen; CK Project*
Environmen-   *No. 2331.* Letter, with attachments, to Isaac Berzin (GreenFuel Technologies Corporation).
tal 2004   September 29, 2004.

ClimateMaster 2004

ClimateMaster. *Tranquility 27™ TT Series: Geothermal Heating and Cooling System: Sizes 026-064*. Unit Specifications. Oklahoma City: ClimateMaster, 2004. On the Web at http://www.climatemaster.com/downloads/RP415.pdf.

ClimateMaster 2007

ClimateMaster. *Tranquility Series*. Oklahoma City: ClimateMaster, 2007. On the Web at http://www.climatemaster.com/index/res_tranquility27_page. "Residential Units."

Cole 1953

Sterling Cole. Letter to Congressman John Phillips, May 20, 1953, with cover note from AEC secretary Roy Snapp, July 9, 1953, DOE Archives, Record Group 326. Box 1290. Folder 2.

COS-Trust 2007

Canadian Oil Sands Trust. *2007 Guidance Document*. [Calgary, Alberta]: COS-Trust, January 29, 2007. On the Web at http://www.cos-trust.com/files/investor/pdf/2007/2007_Guidance.pdf.

CPUC 2006

Dorothy Duda. *Opinion Adopting Performance-Based Incentives, An Administrative Structure, And Other Phase One Program Elements For The California Solar Initiative*. Decision Draft Decision Of ALJ Duda (Mailed 7/24/2006) Before The Public Utilities Commission Of The State Of California. Order Instituting Rulemaking Regarding Policies, Procedures and Rules for the California Solar Initiative, the Self-Generation Incentive Program and Other Distributed Generation Issues. Rulemaking 06-03-004 (Filed March 2, 2006). ALJ/DOT/sid. Agenda ID #5846, Ratesetting. Accompanied by a transmittal letter from Angela K. Minkin, dated July 24, 2006. On the Web at http://www.cpuc.ca.gov/EFILE/DD/58361.pdf

Craig 2004

Paul P. Craig. "Rush to Judgment at Yucca Mountain." *Science for Democratic Action*, v.12, no.3 (June 2004). On the Web at http://www.ieer.org/sdafiles/12-3.pdf.

DARPA 2006

Defense Advanced Research Projects Agency. *BAA06-43 Proposer Information Pamphlet (PIP) for Defense Advanced Research Projects Agency (DARPA), Advanced Technology Office (ATO): BioFuels*. [Washington, DC]: Federal Business Opportunities, July 5, 2006. On the Web at http://www.darpa.mil/baa/baa06-43.html or with links at http://www.fbo.gov/spg/ODA/DARPA/CMO/BAA06%2D43/Attachments.html.

Delft Hydraulics 2006

Aljosja Hooijer, Marcel Silvius, Henk Wösten, and Susan Page. *PEAT-CO2: Assessment of CO2 Emissions from Drained Peatlands in SE Asia*. 1st ed. Report R&D projects Q3943, Q3684, Q4142. WL Delft Hydraulics in co-operation with Wetlands International and Alterra. Delft: Delft Hydraulics, December 7, 2006. On the Web at http://www.wetlands.org/getfilefromdb.aspx?ID=b16d46c5-ea7b-469a-a265-408b59aab5d1.

Deller, Makhijani, and Burroughs 2003

Nicole Deller, Arjun Makhijani, and John Burroughs, eds. *Rule of Power or Rule of Law? An Assessment of U.S. Policies and Actions Regarding Security-Related Treaties*. New York: Apex Press, 2003.

DOE 2006

United States. Department of Energy. *New World Record Achieved in Solar Cell Technology: New Solar Cell Breaks the "40 Percent Efficient" Sunlight-to-Electricity Barrier*. Washington, DC: DOE, Office of Public Affairs, December 5, 2006. On the Web at http://www.energy.gov/news/4503.htm.

DOE 2007

United States. Department of Energy. Office of Energy Efficiency and Renewable Energy. Hydrogen, Fuel Cells & Infrastructure Technologies Program. *Multi-Year Research, Development and Demonstration Plan: Planned Program Activities for 2004-2015*. Washington, DC: EERE, 2005, with updates as of April 27, 2007. Links on Web at http://www1.eere.energy.gov/hydrogenandfuelcells/mypp/, to Chapter 3.1 at http://www1.eere.energy.gov/hydrogenandfuelcells/mypp/pdfs/production.pdf.

DOE 2007a   United States. Department of Energy. *DOE Selects 13 Solar Energy Projects for Up to $168 Million in Funding: First Funding Awards for Solar America Initiative to Make Solar Technology Cost-Competitive by 2015.* Washington, DC: DOE, Office of Public Affairs, March 8, 2007. On the Web at http://www.energy.gov/news/4855.htm.

DOE and EPA 2007   United States. Department of Energy. Office of Energy Efficiency and Renewable Energy and United States. Environmental Protection Agency. *Advanced Technologies and Energy Efficiency.* [Washington, DC: DOE and EPA, 2007]. On the Web at http://www.fueleconomy.gov/feg/atv.shtml.

DSIRE 2007   North Carolina Solar Center and the Interstate Renewable Energy Council. *Database of State Incentives for Renewables & Efficiency.* [Raleigh]: NC Solar Center, [2007]. On the Web http://www.dsireusa.org.

DuPont 2006   Dupont and BP. *Biobutanol Fact Sheet.* [Wilmington, DE]: Dupont, [2006]. On the Web at http://www2.dupont.com/Biofuels/en_US/facts/BiobutanolFactSheet.html.

DuPont 2006b   Dupont. *Progress Report Data Summary.* [Wilmington, DE]: Dupont, 2006. On the Web at http://www2.dupont.com/Sustainability/en_US/Performance_Reporting/data_summary.html#4. From link at *Performance and Reporting: Safety, Health and Environmental Progress.* On the Web at http://www2.dupont.com/Sustainability/en_US/Performance_Reporting/progress.html.

earth2tech 2007   Katie Fehrenbacher. *10 Questions for Nanosolar CEO Martin Roscheisen.* [San Francisco]: GigaOmniMedia, posted July 30, 2007. On the Web at http://earth2tech.com/2007/07/30/10-questions-for-nanosolar-ceo-martin-roscheisen.

Earth Policy 2001   Earth Policy Institute. *Land Area Consumed by the Car in Selected Countries.* Calculations by Janet Larsen. [Washington, DC]: The Institute, [2001]. On the Web at http://www.earth-policy.org/Alerts/Alert12_data2.htm. Title from Web page heading. Linked from article by Lester R. Brown, "Paving the Planet: Cars and Crops Competing for Land," February 14, 2001. On the Web at http://www.earth-policy.org/Alerts/Alert12.htm.

Eberhard and Tarpenning 2006   Martin Eberhard and Marc Tarpenning. *The 21st Century Electric Car.* [San Carlos, CA]: Tesla Motors Inc., October 6, 2006. On the Web at http://www.teslamotors.com/display_data/twentyfirstcenturycar.pdf.

EERE 2004   United States. Department of Energy. Office of Energy Efficiency and Renewable Energy. Building Technologies Program. *Buildings Database.* [Golden, CO]: DOE, EERE, last updated: 08/31/2004. On the Web at http://www.eere.energy.gov/buildings/database/index.cfm.

Glossary of Energy-Related Terms, at http://www.eere.energy.gov/consumer/information_resources/index.cfm/mytopic=60001.

EERE 2006   United States. Department of Energy. Office of Energy Efficiency and Renewable Energy.
                 2006 Buildings Energy Data Book. Prepared for the Buildings Technologies Program and
                 Office of Planning, Budget, and Analysis, Energy Efficiency and Renewable Energy, U.S.
                 Department of Energy, by D&R International, Ltd., under contract to Oak Ridge National
                 Laboratory. [Washington, DC]: DOE, September 2006. On the Web at http://
                 buildingsdatabook.eren.doe.gov/docs/2006bedb-1204.pdf or http://buildingsdatabook.eren.
                 doe.gov/docs/2006bedb-1204.xls.

                 Table 1.2.3 2004 Residential Energy End-Use Splits, by Fuel Type (quads), at http://
                 buildingsdatabook.eren.doe.gov/docs/1.2.3.xls

                 Table 1.3.3 2004 Commercial Energy End-Use Splits, by Fuel Type (quads), at http://build-
                 ingsdatabook.eren.doe.gov/docs/1.3.3.xls

                 Table 2.1.1 Total Number of Households and Buildings, Floorspace, and Household Size, by
                 Year, at http://buildingsdatabook.eere.energy.gov/docs/2.1.1.xls

                 Table 2.2.1 Total Commercial Floorspace and Number of Buildings, by Year, at http://
                 buildingsdatabook.eere.energy.gov/docs/2.2.1.xls

EERE 2006b   United States. Department of Energy. Office of Energy Efficiency and Renewable Energy.
                 Building Technologies Program: Appliances and Commercial Equipment Standards. *History
                 of Federal Appliance Standards.* [Golden, CO]: DOE, EERE, last updated 08/10/2006. On
                 the Web at http://www.eere.energy.gov/buildings/appliance_standards.

EERE Cam-   United States. Department of Energy. Office of Energy Efficiency and Renewable Energy.
bria 2002   Building Technologies Program. *Buildings Database: DEP Cambria: Energy.* [Golden, CO]:
                 DOE, EERE, last updated 04/29/2002. On the Web at http://www.eere.energy.gov/
                 buildings/database/energy.cfm?ProjectID=47.

EERE Durant   United States. Department of Energy. Office of Energy Efficiency and Renewable Energy.
2007   Building Technologies Program. *Buildings Database: Durant Road Middle School: Energy.*
                 [Golden, CO]: DOE, EERE, last updated 02/05/2007. On the Web at http://www.eere.
                 energy.gov/buildings/database/energy.cfm?ProjectID=46.

EERE Ha-   United States. Department of Energy. Office of Energy Efficiency and Renewable Energy.
nover 2002   Building Technologies Program. *Buildings Database: Hanover House: Energy.* [Golden, CO]:
                 DOE, EERE, last updated 06/11/2002. On the Web at http://www.eere.energy.gov/
                 buildings/database/energy.cfm?ProjectID=49.

EERE Solar   United States. Department of Energy. Office of Energy Efficiency and Renewable Energy.
2007   Solar Energy Technologies Program. *Solar FAQs – Photovoltaics – The Basics.* [Golden,
                 CO]: DOE, EERE, last updated 02/08/2007. On the Web at http://www.eere.energy.gov/
                 solar/cfm/faqs/third_level.cfm/name=Photovoltaics/cat=The%20Basics.

EERE Takoma   United States. Department of Energy. Office of Energy Efficiency and Renewable Energy.
2003   Building Technologies Program. *Buildings Database: Takoma Village Cohousing.* [Golden,
                 CO]: DOE, EERE, last updated 04/29/2002 [2003]. Overview on the Web at http://www.
                 eere.energy.gov/buildings/database/overview.cfm?projectid=70.

                 Energy, at http://www.eere.energy.gov/buildings/database/energy.cfm?ProjectID=49.

                 Finance, at http://www.eere.energy.gov/buildings/database/finance.cfm?ProjectID=70.

EIA 1986   United States. Department of Energy. Energy Information Administration. *Analysis of
                 Nuclear Power Plant Construction Costs.* DOE/EIA-0485. Washington, DC: DOE, EIA, Office
                 of Coal, Nuclear, Electric and Alternate Fuels, 1986. DE86008419.

EIA 1995 Coal	United States. Department of Energy. Energy Information Administration. *Coal Data: A Reference*. DOE/EIA-0064(93). Washington, DC: EIA, Office of Coal, Nuclear, Electric and Alternate Fuels, February 1995. On the Web at http://tonto.eia.doe.gov/FTPROOT/coal/006493.pdf.
EIA 1995 Renewables	United States. Department of Energy. Energy Information Administration. *Renewable Energy Annual 1995*. DOE/EIA-0603(95). Washington, DC: EIA, Office of Coal, Nuclear, Electric and Alternate Fuels, December 1995. On the Web at http://tonto.eia.doe.gov/FTPROOT/renewables/060395.pdf.

Parts also on the Web at http://www.eia.doe.gov/cneaf/solar.renewables/renewable.energy.annual/backgrnd/tablecon.htm.

EIA 2001	United States. Department of Energy. Energy Information Administration. *Energy Price Impacts on the U.S. Economy*. Washington, DC: EIA, April 2001. On the Web at http://www.eia.doe.gov/oiaf/economy/energy_price.html, with link to http://www.eia.doe.gov/oiaf/economy/images/figure_1.gif and http://www.eia.doe.gov/oiaf/economy/images/figure_2.gif and whole report at http://www.eia.doe.gov/oiaf/economy/energy_price.pdf.
EIA 2007	United States. Department of Energy. Energy Information Administration. *Energy Expenditure Share of the Economy* [Excel file <Energy Exp Share of GDP_0521.xls>] received by email from Kay Smith, May 21, 2007-07
EIA AEO 2006	United States. Department of Energy. Energy Information Administration. *Annual Energy Outlook 2006 with Projections to 2030*. Report no. DOE/EIA-0383(2006). Washington, DC: EIA, Office of Integrated Analysis and Forecasting, February 2006. Full report on the Web at http://www.eia.doe.gov/oiaf/archive/aeo06/pdf/0383(2006).pdf. Main links at http://www.eia.doe.gov/oiaf/archive/aeo06/index.html.

Reference Case Tables (which also appear as Appendix A of the full report.)

Table 1 Total Energy Supply and Disposition Summary, at http://www.eia.doe.gov/oiaf/archive/aeo06/excel/aeotab_1.xls.

Table 2 Energy Consumption by Sector and Source. On the Web at http://www.eia.doe.gov/oiaf/archive/aeo06/excel/aeotab_2.xls.

Table 7 Transportation Sector Key Indicators and Delivered Energy Consumption. On the Web at http://www.eia.doe.gov/oiaf/archive/aeo06/excel/aeotab_7.xls.

Table 16 Renewable Energy Generating Capacity and Generation. On the Web at http://www.eia.doe.gov/oiaf/archive/aeo06/excel/aeotab_16.xls.

EIA AEO Assumptions 2006	United States. Department of Energy. Energy Information Administration. *Assumptions to the Annual Energy Outlook 2006 with Projections to 2030*. Report no. DOE/EIA-0445(2006). Washington, DC: DOE, EIA, March 2006. Links on the Web at http://www.eia.doe.gov/oiaf/aeo/assumption/index.html. *Residential Demand Module* is on the Web at http://www.eia.doe.gov/oiaf/aeo/assumption/residential.html.

EIA AER 2006   United States. Department of Energy. Energy Information Administration. *Annual Energy Review 2005*. Report No. DOE/EIA-0384(2005). Washington, DC: DOE, EIA, Office of Energy Markets and End Use, July 2006. Links formerly on the Web at http://www.eia. doe.gov/emeu/aer/contents.html. Full report at http://tonto.eia.doe.gov/FTPROOT/ multifuel/038405.pdf.

Table 1.3 Energy Consumption by Source, 1949-2005. Table 2.1a Energy Consumption by Sector, 1949-2005. Table 2.1d Industrial Sector Energy Consumption, 1949–2005.

Table 2.1a Energy Consumption by Sector, Selected Years, 1949-2005

Table 2.1d Industrial Sector Energy Consumption, Selected Years, 1949-2005

Table 5.1 Petroleum Overview, 1949-2005 (Thousand Barrels per Day

Table 5.21 Crude Oil Refiner Acquisition Costs, 1968-2005.

Table 6.5 Natural Gas Consumption by Sector, 1949-2005.

Table 6.8 Natural Gas Prices by Sector, 1967-2005.

Table 8.2a Electricity Net Generation: Total (All Sectors), 1949-2005.

Table 8.4a Consumption for Electricity Generation by Energy Source: Total (All Sectors), 1949-2005.

Table 8.9 Electricity End Use, 1949-2005.

Table 8.10. Average Retail Prices of Electricity, 1960-2005.

History-Petroleum. Energy in the United States: 1635-2000: Petroleum. On the Web at http://www.eia.doe.gov/emeu/aer/eh/petro.html, from a link at http://www.eia.doe. gov/emeu/aer/eh/frame.html.

EIA AER 2007   United States. Department of Energy. Energy Information Administration. *Annual Energy Review 2006*. Report No. DOE/EIA-0384(2006). Washington, DC: DOE, EIA, Office of Energy Markets and End Use, June 2007. Links on the Web at http://www.eia.doe.gov/ emeu/aer/contents.html. Full report at http://www.eia.doe.gov/emeu/aer/pdf/aer.pdf.

Table 1.1 Energy Overview, 1949-2006 (Billion Btu), at http://www.eia.doe.gov/aer/ txt/stb0101.xls, from a link at http://www.eia.doe.gov/emeu/aer/overview.html. Viewed 8-2007.

EIA CABS 2005 US   United States. Department of Energy. Energy Information Administration. *Country Analysis Briefs: United States: Oil*. [Washington, DC]: DOE, EIA, November 2005. On the Web at http://www.eia.doe.gov/emeu/cabs/Usa/Oil.html. Web banner: United States Energy Data, Statistics and Analysis - Oil, Gas, Electricity, Coal. Full report at http://www. eia.doe.gov/emeu/cabs/Usa/pdf.pdf

EIA CABS 2006 Oil Prices   United States. Department of Energy. Energy Information Administration. *Country Analysis Briefs: Annual Oil Market Chronology*. [Washington, DC]: DOE, EIA, [Office of Energy Markets and End Use], May 2006. On the Web at http://www.eia.doe.gov/emeu/ cabs/AOMC/Overview.html. Web banner: Annual Oil Market Chronology Energy Data, Statistics and Analysis - Oil, Gas, Electricity, Coal.

EIA Electric
Power 2006

United States. Department of Energy. Energy Information Administration. *Electric Power Annual 2005*. Revised data. DOE/EIA-0348(2005). Washington, DC: EIA, Office of Coal, Nuclear, Electric and Alternate Fuels, November 2006. On the Web at http://www.eia.doe.gov/cneaf/electricity/epa/epa.pdf, from a link at http://www.eia.doe.gov/cneaf/electricity/epa/epa_sum.html.

Table 1.1. Net Generation by Energy Source by Type of Producer, 1994 through 2005, at http://www.eia.doe.gov/cneaf/electricity/epa/epaxlfile1_1.xls.

Table 2.8. Fuel Switching Capacity of Generators Reporting Natural Gas as the Primary Fuel, by Producer Type, 2005, at http://www.eia.doe.gov/cneaf/electricity/epa/epaxlfile2_8.xls.

EIA EPM
2007-08

United States. Department of Energy. Energy Information Administration. *Electric Power Monthly*. August 2007 Edition. DOE/EIA-0226 (2007/08). Washington, DC: EIA, Office of Coal, Nuclear, Electric and Alternate Fuels, data for May 2007, released: August 15, 2007. On the Web http://www.eia.doe.gov/cneaf/electricity/epm/epm_sum.html.

Table 5.6.A Average Retail Price of Electricity to Ultimate Customers by End-Use Sector, by State, at http://www.eia.doe.gov/cneaf/electricity/epm/table5_6_a.html.

EIA Gas
Primer 2006

United States. Department of Energy. Energy Information Administration. *A Primer on Gasoline*. DOE/EIA-X040. EIA Brochures. Washington, DC: EIA, May 2006. On the Web at http://www.eia.doe.gov/bookshelf/brochures/gasolinepricesprimer/eia1_2005primerM.html.

EIA GHG
2005

United States. Department of Energy. Energy Information Administration. *Emissions of Greenhouse Gases in the United States 2004*. DOE/EIA-0573(2004). Washington, DC: DOE, EIA, Office of Integrated Analysis and Forecasting, December 2005. On the Web at http://www.eia.doe.gov/oiaf/1605/gg05rpt/index.html.

EIA GHG
2005 Docs

United States. Department of Energy. Energy Information Administration. *Documentation for Emissions of Greenhouse Gases in the United States 2004*. DOE/EIA-0638 (2004). Washington, DC: DOE, EIA, Office of Integrated Analysis and Forecasting, December 2006. On the Web at http://www.eia.doe.gov/oiaf/1605/ggrpt/documentation/pdf/0638(2004).pdf.

EIA GHG
2006

United States. Department of Energy. Energy Information Administration. *Emissions of Greenhouse Gases in the United States 2005*. DOE/EIA-0573(2005). Washington, DC: DOE, EIA, Office of Integrated Analysis and Forecasting, November 2006. On the Web at http://www.eia.doe.gov/oiaf/1605/ggrpt/index.html. Full report at http://www.eia.doe.gov/oiaf/1605/ggrpt/pdf/057305.pdf.

Special topics, at http://www.eia.doe.gov/oiaf/1605/ggrpt/stopics.html.

EIA IEA 2006   United States. Department of Energy. Energy Information Administration. *International Energy Annual 2004.* [DOE/EIA-0219(2004)]. [Washington, DC]: DOE, EIA, International Energy Data and Analysis, May-July 2006. Links on the Web at http://www.eia.doe. gov/emeu/iea/contents.html.

World Energy Overview: 1994-2004, at http://www.eia.doe.gov/iea/overview.html.

Table E.1p World Energy Intensity – Total Primary Energy Consumption per Dollar of Gross Domestic Product Using Purchasing Power Parities, 1980-2004, at http://www.eia.doe. gov/pub/international/iealf/tablee1p.xls.

Table H.1co2 World Carbon Dioxide Emissions from the Consumption and Flaring of Fossil Fuels, 1980-2004, at http://www.eia.doe.gov/pub/international/iealf//tableh1co2.xls.

EIA IEO 2006   United States. Department of Energy. Energy Information Administration. *International Energy Outlook 2006.* [DOE/EIA-0484(2006)]. Washington, DC: DOE, EIA, Office of Integrated Analysis and Forecasting, June 2006. Links on the Web at http://www.eia. doe.gov/oiaf/archive/ieo06/index.html. Full report at http://www.eia.doe.gov/oiaf/ieo/ pdf/0484(2006).pdf.

EIA IEO 2007   United States. Department of Energy. Energy Information Administration. *International Energy Outlook 2007.* DOE/EIA-0484(2007). Washington, DC: DOE, EIA, Office of Integrated Analysis and Forecasting, May 2007. Links on the Web at http://www.eia.doe.gov/oiaf/ ieo/index.html. Full report at http://www.eia.doe.gov/oiaf/ieo/pdf/0484(2007).pdf.

Figure 17 data at http://www.eia.doe.gov/oiaf/ieo/excel/figure_17data.xls.

EIA Kyoto   United States. Department of Energy. Energy Information Administration. *Impacts of the*
1998   *Kyoto Protocol on U.S. Energy Markets and Economic Activity.* SR/OIAF/98-03. Washington, DC: DOE, EIA, Office of Integrated Analysis and Forecasting, October 1998. On the Web at http://www.eia.doe.gov/oiaf/kyoto/pdf/sroiaf9803.pdf.

EIA Petro-   United States. Department of Energy. Energy Information Administration. *Petroleum Navi-*
leum 2007   *gator: U.S. Net Imports by Country: Total Crude Oil and Products.* [Washington, DC]; EIA, last updated 03/15/2007. On the Web at http://tonto.eia.doe.gov/dnav/pet/pet_move_ neti_a_ep00_IMN_mbblpd_a.htm, with link from http://www.eia.doe.gov/oil_gas/ petroleum/info_glance/petroleum.html.

EIA Petro-   United States. Department of Energy. Energy Information Administration. *Petroleum*
leum Persian   *Navigator: U.S. Crude Oil and Petroleum Products Imports from Persian Gulf Countries*
Gulf 2007   *(Thousand Barrels).* [Washington, DC]: EIA, last updated 10/02/2006. On the Web at http://tonto.eia.doe.gov/dnav/pet/hist/mttimuspg1A.htm.

EIA Spot   United States. Department of Energy. Energy Information Administration. *Petroleum*
Prices   *Navigator: Spot Prices: (Crude Oil in Dollars per Barrel, Products in Cents per Gallon).* [Washington, DC]: EIA. On the Web at http://tonto.eia.doe.gov/dnav/pet/pet_pri_ spt_s1_d.htm. Viewed July 3, 2007.

EIA STEO   United States. Department of Energy. Energy Information Administration. *STEO Supple-*
2006   *ment: Why Are Oil Prices So High?* [Washington, DC: EIA, Office of Energy Markets and End Use, 2006]. On the Web at http://www.eia.doe.gov/emeu/steo/pub/special/ high-oil-price.html.

Energy Policy Act of 2005	"Energy Policy Act of 2005." Public Law 109-58, 109th Congress (August 8, 2005). 119 *STAT*. 594-1143. On the Web at http://frwebgate.access.gpo.gov/cgi-bin/useftp. cgi?IPaddress=162.140.64.21&filename=publ058.pdf&directory=/diskb/wais/data/109_cong_public_laws.
Energy Services 2003	Western Area Power Administration. "Wind Plus Compressed Air Equals Efficient Energy Storage in Iowa Proposal." *Energy Services Bulletin*, v.22, no.4 (August 2003). On the Web at http://www.wapa.gov/es/pubs/ESB/2003/03Aug/esb084.htm.
EnerNex 2006	EnerNex Corporation. *Final Report - 2006 Minnesota Wind Integration Study, Volume I.* Prepared by: EnerNex Corporation, in collaboration with The Midwest Independent System Operator. Knoxville, TN: EnerNex, November 30, 2006. On the Web at http://www.puc.state.mn.us/docs/windrpt_vol%201.pdf. "Prepared for: The Minnesota Public Utilities Commission."
Environmental Defense 1999	Environmental Defense. *Environmental Justice in Los Angeles: A Timeline.* [New York]: ED, posted on 01/01/1999. On the Web at http://www.environmentaldefense.org/article.cfm?ContentID=2816.
EPA 1988	United States. Environmental Protection Agency. *Design Manual: Constructed Wetlands and Aquatic Plant Systems for Municipal Wastewater Treatment.* EPA/625/1-88/022. Cincinnati, OH: EPA, Office of Research and Development, Center for Environmental Research Information, September 1988. On the Web at http://www.epa.gov/owow/wetlands/pdf/design.pdf.
EPA 1993	United States. Environmental Protection Agency. Science Advisory Board. *Review of the Release of Carbon-14 in Gaseous from High-Level Waste Disposal.* EPA-SAB-RAC-93-010. Washington, DC: U.S. Environmental Protection Agency, 1993.
EPA Fact Sheet SO$_2$	United States. Environmental Protection Agency. *Acid Rain Program SO$_2$ Allowances Fact Sheet.* Clean Air Markets. [Washington]: EPA, last updated February 1, 2007. On the Web at http://epa.gov/airmarkets/trading/factsheet.html#what.
EPA GHG 2006	United States. Environmental Protection Agency. *Global Greenhouse Gas Data.* [Washington, DC]: EPA, updated October 19, 2006. On the Web at http://www.epa.gov/climatechange/emissions/globalghg.html. At head of title: Climate Change – Greenhouse Gas Emissions.
EPA GHG 2007	United States. Environmental Protection Agency. *Inventory of U.S. Greenhouse Gas Emissions and Sinks: 1990 – 2005.* EPA 430-R-07-002. Washington, DC: EPA, Office of Atmospheric Programs, April 15, 2007. On the Web at http://epa.gov/climatechange/emissions/downloads06/07CR.pdf.
EPP 1974	Energy Policy Project of the Ford Foundation. *A Time to Choose: America's Energy Future: Final Report by the Energy Policy Project of the Ford Foundation.* Cambridge, MA: Ballinger, 1974. On the Web at http://www.fordfound.org/eLibrary/documents/0159/toc.cfm.
EPRI 2005	Electric Power Research Institute. *Wave Energy Potential Warrants Further Research and Development, Says EPRI.* News. Palo Alto, CA: EPRI, February 3, 2005. On the Web at http://www.epri.com/corporate/discover_epri/news/2005/020205_WaveEnergy.html.
EPRI 2005b	Electric Power Research Institute. Global Climate Change Research Area, Environment Sector, Technology Group. *Upstream and Downstream Approaches to Carbon Dioxide Regulation.* Climate Brief. Palo Alto, CA: EPRI, 2005. On the Web at http://www.epriweb.com/public/000000000001007762.pdf.

ERT 2005 — Environmental Resources Trust. *Final Report of the Zero Energy Homes for Chicago EcoPower® Project.* Prepared ... for The Illinois Department of Commerce and Economic Opportunity and the Department of Energy – Chicago Regional Office. Washington, DC: ERT, December, 2005. On the Web at http://www.carbonfund.org/Documents/ERT%20Final%20Report.pdf.

Escobar 2001 — Pepe Escobar. "Global Economy: The Roving Eye: Oil's Slippery Slope." *Asia Times Online,* Aug 24, 2004. On the Web at http://www.atimes.com/atimes/Global_Economy/FH24Dj01.html.

Eskom 2005 — Eskom News. "SA Researcher Makes Solar Breakthrough" June 10, 2005. On the Web at http://www.eskom.co.za/live/content.php?Item_ID=702.

Ethanol Market 2007 — Ethanol Market. *United States Fuel Ethanol Industry Sets New All Time High Monthly Production Total.* [Lexington, KY]: Ethanol Market, January 1, 2007. On the Web at http://www.ethanolmarket.com/PressReleaseEthanolMarket010107.html.

European Commission 2000 — European Commission. *Future flight.* European Commission. *Future flight.* [Brussels: Directorate-General for Research], 11-12-2000. Part of the "Competitive and Sustainable Growth (GROWTH) Programme." On the Web at http://ec.europa.eu/research/growth/gcc/projects/in-action-cryoplane.html. Viewed November 2006.

European Parliament 2007 — Kristof Geeraerts. *Limiting Global Climate Change to 2°C: The Way Ahead for 2020 and Beyond.* (IP/A/ENVI/FWC/2006-172/Lot 1/C1/SC7). Report. IP/A/ENVI/WS/2007-8. PE 385.653. Brussels: European Parliament, Policy Department, Economic and Scientific Policy, July 2007. On the Web at http://www.ieep.eu/events/28June-minutes.pdf.

Evergreen Solar 2006 — Evergreen Solar. *Technology: String Ribbon.* Marlboro, MA: ES, [2006]. On the Web at http://www.evergreensolar.com/technology/index.html.

Evergreen Solar 2006b — Evergreen Solar. *Evergreen Solar Announces EverQ Signing of Previously Announced Polysilicon Supply Agreement: Expected to Produce 300 Megawatts of Solar Modules by 2010.* Marlboro, MA: ES, Oct. 2, 2006. On the Web at http://phx.corporate-ir.net/phoenix.zhtml?c=123321&p=irol-newsArticle&ID=910841&highlight. `

Evergreen Solar 2006c — Evergreen Solar. *German Regulatory Authorities Approve EverQ Partnership Agreement.* Marlboro, MA: ES, December 29, 2006. On the Web at http://www.evergreensolar.com/app/en/company/press/pressreleases/item/82.

EVHA 2007 — NAHB Research Center. *Energy Value Housing Award. 2007 EVHA Winners.* [Upper Marlboro, MD]: NAHBRC, 2007. On the Web at http://www.nahbrc.org/evha/winners.html. Viewed August 2007.

Fairley 2007 — Peter Fairley. "Thin Film's Time in the Sun: First Solar's Thin-film Technology is Now Challenging Silicon Panels at Large-scale Solar-power Facilities." Technology Review, July 27, 2007. On the Web at http://www.technologyreview.com/Biztech/19095/?a=f.

Farrell et al. 2006 — Alexander E. Farrell, Richard J. Plevin, Brian T. Turner, Andrew D. Jones, Michael O'Hare, and Daniel M. Kammen. "Ethanol Can Contribute To Energy and Environmental Goals." *Science,* v.311 (27 January 2006), pages 506-508. Erratum published 23 June 2006.

Farrell et al. 2006b — Alexander E. Farrell, Richard J. Plevin, Brian T. Turner, Andrew D. Jones, Michael O'Hare, and Daniel M. Kammen. "Supporting Online Material for: Ethanol Can Contribute To Energy and Environmental Goals." Corrected 27 January 2006; revised version includes table of contents and references. Main article *Science,* v.311 (27 January 2006) pages 506-508. On the Web at http://www.sciencemag.org/cgi/data/311/5760/506/DC1/2.

Feldt 2006 — Richard M. Feldt. *Pacific Growth Equities – Technology Conference, November 8-9, 2006.* [Marlboro, MA]: Evergreen Solar, 2006. On the Web at http://library.corporate-ir.net/library/12/123/123321/items/220495/ESLRPresentationNovember2006.pdf.

First Solar 2007   First Solar, Inc. *United States Securities and Exchange Commission: Form 8-K: Current Report: Pursuant to Section 13 or 15(d) of the Securities Exchange Act of 193. Date of Report (Date of earliest event reported): February 13, 2007.* On the Web at http://files. shareholder.com/downloads/FSLR/99721133x0xS950123%2D07%2D1971/1274494/filing. pdf.

Ford 2004   Ford Motor Company. *Ford Moves Forward with Hydrogen Engine Research.* Dearborn, MI: Ford, September 13, 2004. On the Web at http://media.ford.com/newsroom/feature_ display.cfm?release=18794.

France 2003   France. Ministre déléguée à l'industrie. *Une canicule exceptionnelle pendant l'été 2003.* [Paris]: Ministère de l'Économie, des Finances et de l'Industrie, 28 novembre 2003. On the Web at http://www.industrie.gouv.fr/infopres/presse/Fiche1.pdf.

Freeman 2007, forthcoming   S. David Freeman. *Winning Our Energy Independence.* Layton, UT: Gibbs Smith, forthcoming October 2007.

Fritsche and Lim 2006   Uwe R. Fritsche, with support from Sui-San Lim. *Comparison of Greenhouse-Gas Emissions and Abatement Cost of Nuclear and Alternative Energy Options from a Life-Cycle Perspective.* Updated version. Darmstadt: Öko-Institut e.V., January 2006. On the Web at http://www.newnuclearnothanks.freeola.com/greenhouse_gas_comparison.pdf.

Fthenakis and Kim 2006   Vasilis M. Fthenakis and Hyung Chul Kim. *CdTe Photovoltaics: Life Cycle Environmental Profile and Comparisons.* Presented at the European Material Research Society Meeting, Symposium O, Nice, France, May 29-June 2, 2006 On the Web at http://www.nrel.gov/pv/thin_film/docs/fthenakis_france_6_6_06.pdf.

G8 Climate Declaration 2007   G8 Summit 2007 Heiligendamm. *Growth and Responsibility in the World Economy: Summit Declaration (7 June 2007).* [Heiligendamm: The Summit, 2007]. On the Web at http://www.whitehouse.gov/g8/2007/g8agenda.pdf.

Gain 2006   Bruce Gain. "Road Testing BMW's Hydrogen 7." *Wired*, 11-13-06. On the Web at http://www.wired.com/cars/energy/news/2006/11/72100.

Gates 2007   Dominic Gates. "Boeing Rolls Out Dreamliner for the World." *Seattle Times*, July 9, 2007.

Georgia IRP 2007   Georgia Power Company. *2007 Integrated Resource Plan: Main Document.* [Atlanta]: Georgia Power, 2007. On the Web at http://www.psc.state.ga.us/facts/docftp. asp?txtdocname=99381.

Georgia Tech 2006   T.J. Becker. *Hydrogen Fuel Cells Power Unmanned Aerial Vehicle: Georgia Tech Researchers Conduct Flight Tests Using Compressed Hydrogen.* Atlanta: Research News & Publications Office, Georgia Institute of Technology, August 28, 2006. On the Web at http://www.gatech.edu/news-room/release.php?id=1111.

Ghirardi and Seibert 2003   Maria L. Ghirardi and Michael Seibert. *Algal Hydrogen Photoproduction.* [Presented at the] Program Review Meeting, Berkeley, CA, May 19-22, 2003. Golden, CO: National Renewable Energy Laboratory, 2003. On the Web at http://www1.eere.energy.gov/hydrogenandfuelcells/pdfs/merit03/42_nrel_maria_ghirardi.pdf.

GNEP   Global Nuclear Energy Partnership Web site. On the Web at http://www.gnep.energy.gov/.

Godoy 2006   Julio Godoy. "European Heat Wave Shows Limits of Nuclear Energy." *OneWorld.net*, July 28, 2006. On the Web at http://www.commondreams.org/headlines06/0728-06.htm.

Google 2007   RechargeIT.org. *Recharge a Car, Recharge the Grid, Recharge the Planet.* [Mountain View, CA]: Google, 2007. On the Web at http://www.google.org/recharge. Viewed July 11, 2007.

Google Blog 2006   Robyn Beavers. *Corporate Solar is Coming.* Posted by Robyn Beavers, [Google], 10/16/2006. On the Web at http://googleblog.blogspot.com/2006/10/corporate-solar-is-coming.html. Viewed July 2007.

Gore 2007   Al Gore. "Moving Beyond Kyoto," *New York Times*, July 1, 2007.

Greene et al.   Nathanael Greene and contributing authors, Fuat E. Celik, et al. *Growing Energy: How*
2004   *Biofuels Can Help End America's Oil Dependence*. New York: Natural Resources Defense
                   Council, December 2004. On the Web at http://www.nrdc.org/air/energy/biofuels/
                   biofuels.pdf.

GWEC 2007   Global Wind Energy Council. *Global Wind Energy Markets Continue To Boom – 2006 An-*
                   other Record Year: Industry Delivered 32% of Annual Market Growth Despite Supply Chain
                   Difficulties. Press Release. Brussels: GWEC, 2 February 2007. On the Web at http://www.
                   gwec.net/uploads/media/07-02_PR_Global_Statistics_2006.pdf.

Hadley   Hadley Centre. *Climate Change and the Greenhouse Effect: A Briefing From the Hadley*
Centre 2005   *Centre*. Exeter: Met Office, December 2005. On the Web at http://www.metoffice.gov.
                   uk/research/hadleycentre/pubs/brochures/2005/climate_greenhouse.pdf.

Herrmann,   Ulf Herrmann, Michael Geyer, and Dave Kearney. *Overview on Thermal Storage Systems.*
Geyer, and   [Presented at the] Workshop on Thermal Storage for Trough Power Systems, February 20
Kearney   - 21, 2002. [Köln]: FLABEG Solar International GmbH, 2002. On the Web at http://www.
2002   nrel.gov/csp/troughnet/pdfs/uh_storage_overview_ws030320.pdf.

Hong and   B.D. Hong and E. R. Slatick. "Carbon Dioxide Emission Factors for Coal." Originally
Slatick 1994   published in, *Quarterly Coal Report*, January-April 1994, pages 1-8. DOE/EIA-0121(94/Q1).
                   Washington, DC, Energy Information Administration August 1994. On the Web at http://
                   www.eia.doe.gov/cneaf/coal/quarterly/co2_article/co2.html.

Howdeshell   Kembra L. Howdeshell, Paul H. Peterman, Barbara M. Judy, Julia A. Taylor, Carl E. Orazio,
et al. 2003   Rachel L. Ruhlen, Frederick S. vom Saal, and Wade V. Welshons. "Bisphenol A Is Released
                   from Used Polycarbonate Animal Cages into Water at Room Temperature." *Environmen-*
                   tal Health Perspectives, v.111, no.9 (July 2003). On the Web at http://www.ehponline.
                   org/members/2003/5993/5993.pdf.

Hu and   Patricia Hu and Jenny Young. *1990 NPTS Databook Volumes I and II: Based on Data from*
Young 1994   *the 1990 Nationwide Personal Transportation Survey (NPTS)*. [Washington, DC]: U.S.
                   Department of Transportation, Federal Highway Administration, October 1994. On the
                   Web at http://npts.ornl.gov/npts/1990/pdf/nptsall.pdf, with links from http://npts.ornl.
                   gov/npts/1990/index.html.

Hybridcars.   [Bradley Berman]. *The Model G: Google's Plug-in Hybrid Program*. [no place]: Hybridcars.
com 2007   com, 2007. On the Web at http://www.hybridcars.com/plug-in-hybrids/google-plugin-hy-
                   brid-modelg.html. Viewed July 23, 2007.

Hydrogen   Hydrogen Cars Now. *BMW Hydrogen 7 Luxury Automobile*. [Laguna Beach, CA]: Hydrogen
Cars Now   Cars Now, [2006]. On the Web at http://www.hydrogencarsnow.com/bmw-hydrogen7.htm.
2006
Ice Energy   Ice Energy Web site. [Windsor, CO]: Ice Energy, 2007. On the Web at http://ice-energy.
                   com/. Viewed August 28, 2007.

IEER 1999   Arjun Makhijani. "Some Evidence of Yucca Mountain's Unsuitability as a Repository."
                   Science for Democratic Action, v.7, no.3, May 1999. On the Web at http://www.ieer.
                   org/sdafiles/vol_7/7-3/yucca.html.

Interior 2006   United States. Department of Interior. *Interior Secretary Kempthorne Announces Proposal*
                   to List Polar Bears as Threatened Under Endangered Species Act. News. Washington, DC:
                   Office of the Secretary, December 27, 2006. On the Web at http://www.doi.gov/news/06_
                   News_Releases/061227.html.

Interior 2007   United States. Department of the Interior. Fish and Wildlife Service. "Endangered and Threatened Wildlife and Plants; 12-Month Petition Finding and Proposed Rule To List the Polar Bear (Ursus maritimus) as Threatened Throughout Its Range." 50 CFR Part 17 RIN 1018-AV19. *Federal Register*, v.72, no.5 (January 9, 2007) pages 1064 to 1099,. On the Web at http://a257.g.akamaitech.net/7/257/2422/01jan20071800/edocket.access.gpo. gov/2007/pdf/06-9962.pdf or http://www.epa.gov/fedrgstr/EPA-SPECIES/2007/January/ Day-09/e9962.htm.

IPCC 1999   Joyce E. Penner, David Lister. David J. Griggs, David J. Dokken, and Mack McFarland, editors. *Aviation and the Global Atmosphere: A Special Report of the Intergovernmental Panel on Climate Change*. [Arendal, Norway: UNEP/GRID-Arendal, September 1999]. Links to whole report at http://www.grida.no/climate/ipcc/aviation/index.htm. See "7.8.2. Fuel Composition Effects on Emissions" at http://www.grida.no/climate/ipcc/aviation/110.htm. Also published as a book by Cambridge University Press.

IPCC 2005   Bert Metz, Ogunlade Davidson, Heleen de Coninck, Manuela Loos, and Leo Meyer, editors. *IPCC Special Report on Carbon Dioxide Capture and Storage*. Prepared by Working Group III of the Intergovernmental Panel on Climate Change. Cambridge: Cambridge University Press, 2005. Links at http://www.ipcc.ch/activity/srccs/index.htm.

IPCC 2007   Intergovernmental Panel on Climate Change. *Working Group III Contribution to the Intergovernmental Panel on Climate Change: Fourth Assessment Report: Climate Change 2007: Mitigation of Climate Change: Summary for Policymakers*. Geneva: IPCC Secretariat; WMO; UNEP, 2007. On the Web at http://www.mnp.nl/ipcc/docs/FAR/SPM_%20WGIII_ rev5.pdf.

ISG 2006   Iraq Study Group. *The Iraq Study Group Report*. James A. Baker, III, and Lee H. Hamilton, Co-Chairs. Lawrence S. Eagleburger, Vernon E. Jordan, Jr., Edwin Meese III, Sandra Day O'Connor, Leon E. Panetta, William J. Perry, Charles S. Robb, Alan K. Simpson. [Washington, DC: United States Institute of Peace, December 6, 2006]. On the Web at http://www. usip.org/isg/iraq_study_group_report/report/1206/iraq_study_group_report.pdf.

Ivy 2004   Johanna Ivy. *Summary of Electrolytic Hydrogen Production*. NREL/MP-560-36734. Milestone Completion Report. Golden, CO: National Renewable Energy Laboratory, September 2004. On the Web at http://www.nfpa.org/assets/files/PDF/CodesStandards/ HCGNRELElectrolytichydrogenproduction04-04.pdf.

Karbuz 2007   Sohbet Karbuz. "US Military Oil Pains." *Energy Bulletin*, February 17, 2007. On the Web at http://www.energybulletin.net/26194.html.

Karppi 2002   D. F. Karppi. "Geothermal Energy Comes to Town: LIPA's Clean Energy Initiative Helps Protect Environment, Provide Consumer Savings." *Oyster Bay Enterprise-Pilot*, March 8, 2002. On the Web at http://www.antonnews.com/oysterbayenterprisepilot/2002/03/08/ news.

Keith et al. 2004   David W. Keith, Joseph F. DeCarolis, David C. Denkenberger, Donald H. Lenschow, Sergey L. Malyshev, Stephen Pacala, and Philip J. Rasch. "The influence of large-scale wind power on global climate." *PNAS*, v.101, no.46 (November 16, 2004), pages 16115–16120. On the Web at http://www.pnas.org/cgi/reprint/101/46/16115.pdf.

Kemp 2006   Melissa Kemp (Project Associate, Renovus Energy, Ithaca, NY). Personal communication, with Arjun Makhijani, 20 December 2006.

Kempton and Letendre 1997   Willett Kempton and Steven E. Letendre. "Electric Vehicles as a New Power Source for Electric Utilities." *Transportation Research*, v.2, no.3 (1997), page 157-175. On the Web at http://www.udel.edu/V2G/docs/Kempton-Letendre-97.pdf.

Kennedy et al. 2006

John Kennedy, Andreas Zsiga, Laurie Conheady, and Paul Lund. "Credit Aspects of North American and European Nuclear Power." Reprinted from RatingsDirect. [New York]: Standard & Poor's, January 9, 2006.

Keshner and Arya 2004

M.S. Keshner and R. Arya. *Study of Potential Cost Reductions Resulting from Super-Large-Scale Manufacturing of PV Modules: Final Subcontract Report, 7 August 2003-30 September 2004*. Subcontractor Report. NREL/SR-520-36846. Golden, CO: National Renewable Energy Laboratory, October 2004. On the Web at http://www.nrel.gov/pv/thin_film/docs/keshner.pdf.

Kimball 2004

MEAN Wind Project at Kimball. *Kimball Wind Project Facts*. [Lincoln]: NMPP Energy/Municipal Energy Agency of Nebraska, last updated 08/04. On the Web at http://www.nmppenergy.org/KimballWindProject/Facts.htm. Viewed July 2007.

Kissinger 2007

Henry A. Kissinger. "Stability in Iraq and Beyond." *Washington Post*, January 21, 2007, page B7. Similar version on the Web at http://www.tmsfeatures.com/tmsfeatures/subcategory.jsp?custid=67&catid=1592.

LANL Biography

Los Alamos National Laboratory. *Staff Biographies: Robert Fox Bacher*. On the Web at http://www.lanl.gov/history/people/R_Bacher.shtml. Viewed 1-10-2006.

Light 2003

Peter Light. *Fuel Cell Vehicles as Distributed Power Resources in San Francisco*. 11 November 2003. On the Web at http://www.peterlight.com/energy/rmi/V2G%20for%20SF-03.pdf.

LIPA 2006

Long Island Power Authority. *The Inn at Foxhollow*. [Uniondale, NY]: LIPA, last updated 02/17/2006. On the Web at http://www.lipower.org/cei/geothermal.fox.html.

Makhijani 1990

Arjun Makhijani. *Draft Power in South Asian Foodgrain Production: Analysis of the Problem and Suggestions for Policy*. Takoma Park, MD: Institute for Energy and Environmental Research, September 1990. On the Web at http://www.ieer.org/reports/energy/draftpower.pdf.

Makhijani 1997

Arjun Makhijani. "Achieving Enduring Nuclear Disarmament." *Science for Democratic Action*, v.6, no.4 and v.7, no.1 (October 1998). On the Web at http://www.ieer.org/latest/disarm.html.

Makhijani 2001

Arjun Makhijani. *Plutonium End Game: Managing Global Stocks of Separated Weapons-Usable Commercial and Surplus Nuclear Weapons Plutonium*. Takoma Park, MD: Institute for Energy and Environmental Research, January 22, 2001. On the Web at http://www.ieer.org/reports/pu/peg.pdf.

Makhijani 2004

Arjun Makhijani. "Atomic Myths, Radioactive Realities: Why Nuclear Power Is a Poor Way to Meet Energy Needs." *Journal of Land, Resources, & Environmental Law*, v.24, no.1 (2004), pages 61-72. On the Web at http://www.ieer.org/pubs/atomicmyths.html.

Makhijani and Barczak 2007

Arjun Makhijani and Sara Barczak. *Direct Testimony of Arjun Makhijani, Ph.D. and Sara Barczak on Behalf of Southern Alliance for Clean Energy*. GPSC Docket No. 24505-U. Georgia Power's Integrated Resource Plan. May 4, 2007. On the Web at http://www.psc.state.ga.us/facts/docftp.asp?txtdocname=102016.

Makhijani and Gurney 1995

Arjun Makhijani and Kevin R. Gurney. *Mending the Ozone Hole: Science, Technology and Policy*. Cambridge: MIT Press, 1995.

Makhijani and Lichtenberg 1971	A.B. Makhijani and A.J. Lichtenberg. *An Assessment of Energy and Materials Utilization in the U.S.A.* Memorandum no. ERL-M310 (Revised). Berkeley: Electronics Research Laboratory, College of Engineering, University of California, 22 September 1971, as reprinted in the *Congressional Record*, v.117, part 34 (December 2, 1971 to December 7, 1971), pages 44629-44635. Figures omitted from *Congressional Record.*
Makhijani and Makhijani 2006	Annie Makhijani and Arjun Makhijani. *Low-Carbon Diet Without Nukes in France: An Energy Technology and Policy Case Study on Simultaneous Reduction of Climate Change and Proliferation Risks.* Takoma Park, MD: Institute for Energy and Environmental Research, May 4, 2006. On the Web at http://www.ieer.org/reports/energy/france/lowcarbonreport.pdf.
Makhijani and Poole 1975	Arjun Makhijani, in collaboration with Alan Poole. *Energy and Agriculture in the Third World.* A Report to the Energy Policy Project of the Ford Foundation. Cambridge, MA: Ballinger, 1975. On the Web at http://www.fordfound.org/elibrary/documents/0338/toc.cfm.
Makhijani and Saleska 1999	Arjun Makhijani and Scott Saleska. *The Nuclear Power Deception: U.S. Nuclear Mythology From Electricity "Too Cheap To Meter" To "Inherently Safe" Reactors.* A Report of the Institute for Energy and Environmental Research. New York: Apex Press, 1999.
Makhijani and Smith 2004	Arjun Makhijani and Brice Smith. Costs and Risks of Management and Disposal of Depleted Uranium from the National Enrichment Facility Proposed to be Built in Lea County New Mexico by LES. Takoma Park, MD: Institute for Energy and Environmental Research, November 24, 2004; version for public release redacted March 20, 2007. On the Web at http://www.ieer.org/reports/du/lesrpt.pdf.
Makhijani et al. 2004	Arjun Makhijani, Peter Bickel, Aiyou Chen, and Brice Smith. *Cash Crop on the Wind Farm: A New Mexico Case Study of the Cost, Price, and Value of Wind-Generated Electricity.* Prepared for presentation at the North American Energy Summit Western Governors' Association, Albuquerque, New Mexico, April 15-16, 2004. Takoma Park, MD: Institute for Energy and Environmental Research, April 2004, revised as of February 8, 2005. On the Web at http://www.ieer.org/reports/wind/cashcrop/index.html.
Makhijani, Gunter, and Makhijani 2002	Annie Makhijani, Linda Gunter, and Arjun Makhijani. "COGEMA: Above The Law? Some Facts about the Parent Company of a U.S. Corporation Set to Process Plutonium in South Carolina. *Science for Democratic Action*, v.10, no.3 (May 2002). On the Web at http://www.ieer.org/sdafiles/vol_10/sda10-3.pdf.
Makhijani, Makhijani, and Bickel 1988	Arjun Makhijani, Annie Makhijani, and Amanda Bickel. *Saving Our Skins: Technical Potential and Policies for the Elimination of Ozone-Depleting Chlorine Compounds.* [Takoma Park, MD]: Environmental Policy Institute and Institute for Energy and Environmental Research, 1988.
Malkin 2007	Elizabeth Malkin, with Antonio Betancourt. "Thousands in Mexico City to Protest Rising Food Prices." *New York Times*, February 1, 2007, p. A6.
Meinshausen 2006	Malte Meinshausen. "What Does a 2°C Target Mean for Greenhouse Gas Concentrations? A Brief Analysis Based on Multi-Gas Emission Pathways and Several Climate Sensitivity Uncertainty Estimates." In *Avoiding Dangerous Climate Change*, Hans Joachim Schellnhuber, et al. (eds.), Cambridge: Cambridge University Press, 2006. pages 265 - 280.
Miller 2007	John M. Miller, of Maxwell Technologies, Inc., telephone interview with Arjun Makhijani, January 30 and 31, 2007. Notes of Arjun Makhijani.

Milbrandt 2005	A. Milbrandt. *A Geographic Perspective on the Current Biomass Resource Availability in the United States.* Technical Report. NREL/TP-560-39181. Golden, CO: National Renewable Energy Laboratory, December 2005. On the Web at http://www.nrel.gov/docs/fy06osti/39181.pdf.
MIT 2003	Massachusetts Institute of Technology. *The Future of Nuclear Power: An Interdisciplinary MIT Study.* John Deutch and Ernest J. Moniz (co-chairs) et al. [Cambridge]: MIT, 2003. On the Web at http://web.mit.edu/nuclearpower/pdf/nuclearpower-full.pdf.
MIT 2006	Massachusetts Institute of Technology. *The Future of Geothermal Energy: Impact of Enhanced Geothermal Systems (EGS) on the United States in the 21st Century: An Assessment by an MIT-led Interdisciplinary Panel.* Jefferson W. Tester, Chair. INL/EXT-06-11746. Cambridge: MIT, 2006. On the Web at http://geothermal.inel.gov/publications/future_of_geothermal_energy.pdf.
MIT Cogen 2007	Massachusetts Institute of Technology. *MIT Cogeneration Project.* [Cambridge: MIT, 2007]. On the Web at http://cogen.mit.edu.
MIT LEES 2006	Massachusetts Institute of Technology. Laboratory for Electromagnetic and Electronic Systems. *Laboratory for Electromagnetic and Electronic Systems 2006 Research Report.* [Cambridge]: MIT, LEES, [2006]. On the Web at http://lees-web.mit.edu/lees/Reports/LeesPresReport2006.pdf.
MIT News 2004	Sarah H. Wright. *Up On the Roof, Algae Appetites May Transform Waste Into Energy.* Cambridge: Massachusetts Institute of Technology, News Office, August 9, 2004. On the Web at http://web.mit.edu/newsoffice/2004/algae.html.
Montreal Protocol 2000	*The Montreal Protocol on Substances that Deplete the Ozone Layer as Adjusted and/or Amended in London 1990, Copenhagen 1992, Vienna 1995, Montreal 1997, Beijing 1999.* Ozone Secretariat, United Nations Environment Programme. Nairobi: Secretariat for The Vienna Convention for the Protection of the Ozone Layer & The Montreal Protocol on Substances that Deplete the Ozone Layer, 2000. On the Web at http://hq.unep.org/ozone/Montreal-Protocol/Montreal-Protocol2000.shtml. Text. See overview at http://ozone.unep.org.
Moreland and Collins 1990	Alvin F. Moreland and Bobby R. Collins. *Water Hyacinths (Eichhornia crassipes), Grown In Municipal Wastewater, as a Source of Organic Matter in Rabbit Food.* Final report of a study. [Gainesville: Department of Small Animal Clinical Sciences, College of Veterinary Medicine, University of Florida], June 20, 1990. On the Web at http://www.hydromentia.com/Products-Services/Water-Hyacinth-Scrubber/Product-Documentation/Assets/1990_Moreland-et-al-Water-Hyacinth-Rabbit-Food.pdf.
Mufson and Cho 2007	Steven Mufson and David Cho. "Energy Firm Accepts $45 Billion Takeover: Buyers Made Environmental Pledge." *Washington Post,* February 26, 2007; page A04. On the Web at http://www.washingtonpost.com/wp-dyn/content/article/2007/02/25/AR2007022501520.html.
Murphy et al. 2004	James M. Murphy, David M. H. Sexton, David N. Barnett, Gareth S. Jones, Mark J. Webb, Matthew Collins, and David A. Stainforth. "Quantification of Modelling Uncertainties in a Large Ensemble of Climate Change Simulations." *Nature,* v.430 (12 August 2004), pages 768 – 772.
Murray 1953	Thomas E. Murray. Memorandum to Lewis L. Strauss, September 18, 1953. DOE Archives, Record Group 326, Box 1290, Folder 2.
Musial 2005	Walt Musial. *Offshore Wind Energy Potential for the United States.* Wind Powering America – Annual State Summit, May 19, 2005, Evergreen, Colorado. [Golden, CO: National Renewable Energy Laboratory, 2005]. On the Web at http://www.eere.energy.gov/windandhydro/windpoweringamerica/pdfs/workshops/2005_summit/musial.pdf.

Nadel et al. 2006   Steven Nadel, Andrew deLaski, Maggie Eldridge, and Jim Kliesch. *Leading the Way: Continued Opportunities for New State Appliance and Equipment Efficiency Standards.* Report Number ASAP-6/ACEEE-A062. Washington, DC: American Council for an Energy-Efficient Economy; Boston: Appliance Standards Awareness Project, March 2006. On the Web at http://aceee.org/pubs/a062.pdf?CFID=2147365&CFTOKEN=47763915. Updated from and supersedes Report A051.

Nanosolar Web Site   Nanosolar, Inc. Web site. Palo Alto, CA. On the Web at http://www.nanosolar.com.

Nanosolar Secures $100,000,000 in Funding: Substantial Cleantech Equity Financing for Breakthrough Solar Cell Technology; Cementing Leadership Position Towards Delivering Grid-Parity Peak Power. June 21, 2006. On the Web at http://www.nanosolar.com/pr5.htm.

Nanosolar Selects Manufacturing Sites: Company Secures 647,000 Square Feet of Space in the San Francisco Bay Area and in Germany. December 12, 2006. On the Web at http://www.nanosolar.com/pr7.htm.

History and Milestones. On the Web at http://www.nanosolar.com/history.htm.

Narendran et al. 2005   N. Narendran, Y. Gu, J. P. Freyssinier-Nova, Y. Zhu. "Extracting Phosphor-scattered Photons to Improve White LED Efficiency. Abstract." *Physica status solidi (a)*, v.202, issue 6 (6 Apr 2005), pages R60-R62. On the Web at http://www3.interscience.wiley.com/cgi-bin/abstract/110437401/ABSTRACT?CRETRY=1&SRETRY=0 – with sign in. "Rapid Research Letter."

NAS/NRC 2006   Richard R. Monson (Chair) et al. Health Risks from Exposure to Low Levels of Ionizing Radiation: BEIR VII – Phase 2. Committee to Assess Health Risks from Exposure to Low Levels of Ionizing Radiation, Board on Radiation Effects Research, National Research Council of the National Academies. Washington, DC: National Academies Press, 2006. On the Web at http://books.nap.edu/catalog.php?record_id=11340.

Nature editorial 2007   "Nuclear Test: Japan's Response to an Earthquake Highlights Both the Promise and the Pitfalls of Nuclear Power at a Critical Time for Its Future." Editorial. *Nature*, v.448, issue no.7152 (26 July 2007), page 387.

NCUC 2007   North Carolina Utilities Commission. *In the Matter of: Application for Approval for an Electric Generation Certificate to Construct Two 800 MW State of the Art Coal Units for Cliffside Project in Cleveland/Rutherford Counties.* E-7, Sub 790-Vol. 6, January 19, 2007.

NJC 2007   Nathan Rabinovitch. *How Air Pollution Affects Asthma.* [Denver]: National Jewish Medical and Research Center, 2007. On the Web at http://www.njc.org/disease-info/diseases/asthma/living/environ/pollution.aspx. Taken from an online chat dated June 21, 2000.

NOAA CO_2 Trends   Pieter Tans. *Trends in Atmospheric Carbon Dioxide - Mauna Loa.* [Boulder, CO]: U.S. Department of Commerce, National Oceanic and Atmospheric Administration, Earth System Research Laboratory, Global Monitoring Division, [2007]. On the Web at http://www.esrl.noaa.gov/gmd/ccgg/trends/. Viewed July 5, 2007.

Northbridge 2003   Northbridge Environmental Management Consultants. *Analyzing the Cost of Obtaining LEED Certification.* Arlington, VA: American Chemistry Council, April 16, 2003. On the Web at http://www.greenbuildingsolutions.org/s_greenbuilding/docs/800/776.pdf.

NPT   *Treaty on the Non-proliferation of Nuclear Weapons (NPT).* [New York]: Federation of American Scientists, updated April 27, 1998. Links on the Web at http://www.fas.org/nuke/control/npt/text/index.html. Viewed 2-2-07.

NREL 1998   John Sheehan, Terri Dunahay, John Benemann, and Paul Roessler. *A Look Back at the U.S. Department of Energy's Aquatic Species Program: Biodiesel from Algae.* NREL/TP-580-24190. Golden, CO: National Renewable Energy Laboratory, July 1998. On the Web at http://www1.eere.energy.gov/biomass/pdfs/biodiesel_from_algae.pdf. "Close-out Report."

NREL 2004   National Renewable Energy Laboratory. *NREL and R&D Partners Work to Trim Solar Electricity Costs.* Feature Story. [Golden, CO]: NREL, February 2004. On the Web at http://www.nrel.gov/features/02-04_solar_costs.html. "Contributed by Tom LaRocque."

NREL 2005   National Renewable Energy Laboratory. *Power Technologies Energy Data Book.* 3rd ed. Compiled by J. Aabakken. Technical Report. NREL/TP-620-37930. Golden, CO: NREL, April 2005. On the Web at http://www.nrel.gov/analysis/power_databook_3ed/docs/pdf/complete.pdf. Excel links on the Web at http://www.nrel.gov/analysis/power_databook_3ed/toc.html.

NYC 2007   New York City. Department of City Planning. Transportation Division. *Bicycle & Greenway Planning.* New York: The Department, 2007. On the Web at http://www.nyc.gov/html/dcp/html/transportation/td_projectbicycle.shtml.

NYSERDA   Global Energy Concepts and AWS Scientific, Inc. *Wind Power Project Site Identification & Land Requirements.* A Part of the NYSERDA Wind Power Toolkit. Albany: NYSERDA, June 2005. On the Web at http://text.powernaturally.org/Programs/Wind/toolkit/Land_Requirements.pdf.
2005

Öko-Institut Felix Matthes, Verena Graichen, and Julia Repenning (Öko-Institut); in cooperation with Caroline Doble et al. *The Environmental Effectiveness and Economic Efficiency of the European Union Emissions Trading Scheme: Structural Aspects of Allocation: A Report to WWF.* Brussels: WWF-World Wide Fund For Nature, November 2005. On the Web at http://www.wwf.de/imperia/md/content/klima/2005_11_08_full_final__koinstitut.pdf.
2005

Oliver 2006   Hongyan He Oliver. "Reducing China's Thirst for Foreign Oil: Moving Towards a Less Oil-Dependent Road Transport System." *China Environment Series* 8, 2006. On the Web at http://www.wilsoncenter.org/topics/pubs/CEF_Feature.3.pdf.

O'Neill 2006 Steven M. O'Neill. *2000 U.S. Census Analysis: New York City Vehicle Ownership (By Household) (Revised January 22, 2006).* [New York: Right of Way], January 22, 2006. On the Web at http://www.cars-suck.org/research/2000_5boro_census_revised.html.

ORNL Solar   Oak Ridge National Laboratory. Solar Technologies Program. *Welcome to Solar Technologies at ORNL.* [Oak Ridge, TN]: The Program, last revision June 13, 2007. On the Web at http://www.ornl.gov/sci/solar. Viewed July 18, 2007.
2007

O'Shea 2006 Paul O'Shea. "Tests Confirm Extended Battery Cell Life." *Power Management DesignLine* (November 01, 2006). On the Web at http://www.powermanagementdesignline.com/news/showArticle.jhtml;jsessionid=V21MHN2Z3KWHOQSNDLPCKHSCJUNN2JVN?articleID=193501145.

PACE-Univer- Paper, Allied Industrial, Chemical, and Energy Workers (PACE) International Union and University of Utah. *Exposure Assessment Project at the Paducah Gaseous Diffusion Plant.* Submitted by Paper, Allied Industrial, Chemical, and Energy Workers (PACE) International Union and University of Utah, Division of Radiology, Center for Advanced Medical Technologies, Center Excellence in Nuclear Technology, Engineering and Research. Sponsored by U.S. Department of Energy, Office of Environmental, Safety and Health. December 2000.
sity of Utah
2000

Paley   President's Materials Policy Commission (Paley Commission). *Resources for Freedom.* Washington, DC: Govt. Print. Off., 1952. 5 volumes.
Commission
1952

Parker, Sherwin, and Floyd 1998

Danny Parker, John Sherwin, David B. Floyd. *Measured Energy Savings From Retrofits Installed In Low-Income Housing In A Hot and Humid Climate.* FSEC-PF-339-98. [Orlando]: Florida Solar Energy Center (FSEC), University of Central Florida, 1998. On the Web at http://www.fsec.ucf.edu/en/publications/html/FSEC-PF-339-98/index.htm. "Presented at the 11th Symposium on Improving Building Systems in Hot and Humid Climates, Fort Worth, TX, June 1 and 2, 1998."

Parsons et al. 2006

Brian Parsons, Michael Milligan, J Charles Smith, Edgar DeMeo, Brett Oakleaf, Kenneth Wolf, Matt Schuerger, Robert Zavadil, Mark Ahlstrom, and Dora Yen Nakafuji. *Grid Impacts of Wind Power Variability: Recent Assessments from a Variety of Utilities in the United States.* Presented at the European Wind Energy Conference, Athens, Greece, February 27-March 2, 2006. Conference Paper NREL/CP-500-39955. [Golden, CO]: National Renewable Energy Laboratory, July 2006. On the Web at http://www.nrel.gov/docs/fy06osti/39955.pdf.

Paul 2002

James A. Paul. *Oil in Iraq: the Heart of the Crisis.* New York: Global Policy Forum, December, 2002. On the Web at http://www.globalpolicy.org/security/oil/2002/12heart.htm.

Phoenix Motorcars

Phoenix Motorcars. Web site. [Ontario, CA]: Phoenix Motorcars, 2007. On the Web at http://www.phoenixmotorcars.com. Viewed March 1, 2007. See also http://www.phoenixmotorcars.com/faq/index.html.

Pianin and Goldstein 2001

Eric Pianin and Amy Goldstein. "Bush Drops a Call For Emissions Cuts." *Washington Post,* March 14, 2001. A01.

Pidwirny 2006

Michael Pidwirny. *Fundamentals of Physical Geography.* 2nd ed. (2006). On the Web at http://www.physicalgeography.net/fundamentals/6i.html. Viewed on June 26, 2007.

PowerLight 2002

PowerLight. *United States Navy Deploys Largest Federal Solar Electric System in the Nation: 750 kW Photovoltaic System Installed at Naval Base Coronado, California.* Powerlight Press Release. [Berkeley, CA]: PowerLight, November 08, 2002. On the Web at http://www.powerlight.com/about/press_2002/11_08_02.html.

PRSEA 2003

Potomac Region Solar Energy Association. *13th DC Region Solar Home Tour.* Annapolis, MD: PRSEA, 2003. On the Web at http://www.prsea.org/tour/tour2003.html; cached at http://64.233.169.104/search?q=cache:nVQF0BJLBKMJ:www.prsea.org/tour/tour2003.html+www.prsea.org+tour2003&hl=en&ct=clnk&cd=1&gl=us.

Public Renewables

Public Renewables Partnership. *Wind: Wind Energy Cost Overview.* [no place]: PRP, [2003]. On the Web at http://www.repartners.org/wind/wpcost.htm. Viewed July 2007.

Qusti 2006

Raid Qusti. "GCC to Develop Civilian Nuclear Energy." *Arab News,* 11 December 2006. Reprinted by Saudi-US Relations Information Service in "27th GCC Supreme Council Summit Wrapup." On the Web at http://www.saudi-us-relations.org/articles/2006/ioi/061213-gcc-summit.html.

Renewable Energy Access 2007

Renewable Energy Access. *Iowa to Combine Wind Energy & CAES Technology.* [Peterborough, NH]: REA, January 12, 2007. On the Web at http://www.renewableenergyaccess.com/rea/news/story?id=47096.

Rittenhouse and Carberry 2007

Dawn Rittenhouse and John Carberry. Final summary of telephone conversation with Dawn Rittenhouse and John Carberry, both of DuPont, with Arjun Makhijani, 14 February 2007. Reviewed and corrected by interviewees.

Rosenfeld 2003	Arthur H. Rosenfeld. *Sustainable Development: Reducing Energy Intensity by 2% Per Year: [PowerPoint presentation] Energy PMP, International Seminar on Planetary Emergencies, Erice, Italy 19-8-2003.* [Sacramento: California Energy Commission, 2003]. On the Web http://www.energy.ca.gov/papers/2003-08-19_ROSENFELD.PPT.
Rosenfeld and McAuliffe 2006	Arthur H. Rosenfeld and Pat McAuliffe. *Comments on First 10 Slides of Presentation Entitled Fermi Talks 6-21, 22, 2006.* Rosenfeld and McAuliffe, 6-6-06. [Sacramento: California Energy Commission, 2006]. On the Web at http://www.energy.ca.gov/commission/commissioners/rosenfeld_docs/2006-06-21_FermiSlideNotes.doc.
Rosenthal 2007	Elisabeth Rosenthal. "Once a Dream Fuel, Palm Oil May Be an Eco-Nightmare." *New York Times*, January 31, 2007.
Rundle et al. 2007	Andrew Rundle, Ana V. Diez Roux, Lance M. Freeman, Douglas Miller, Kathryn M. Neckerman, and Christopher C. Weiss. "The Urban Built Environment and Obesity in New York City: A Multilevel Analysis." *American Journal of Health Promotion*, v.21, issue 4 Supplement (March-April 2007), pages 326+. On the Web at http://www.healthpromotionjournal.com.
Runge and Senauer 2007	C. Ford Runge and Benjamin Senauer. "How Biofuels Could Starve the Poor." *Foreign Affairs*, v.86, no.3 (May/June 2007). On the Web at http://www.foreignaffairs.org/20070501faessay86305-p0/c-ford-runge-benjamin-senauer/how-biofuels-could-starve-the-poor.html.
Sachs et al. 2004	H. Sachs, S. Nadel, J. Throne Amann, M. Tuazon, E. Mendelsohn, L. Rainer, G. Todesco, D. Shipley, and M. Adelaar. *Emerging Energy-Saving Technologies and Practices for the Buildings Sector as of 2004.* Report Number A042. Washington, DC: American Council for an Energy-Efficient Economy, October 2004. On the Web at http://aceee.org/pubs/a042full.pdf.
Sandalow 2006	David Sandalow. "Ethanol: Lessons from Brazil." Appendix F of *A High Growth Strategy for Ethanol: The Report of an Aspen Institute Policy Dialogue.* David W. Monsma, Rapporteur; John A. Riggs, Program Executive Director. Washington, DC: Aspen Institute, Program on Energy, the Environment, and the Economy, 2006. On the Web at http://www.aspeninstitute.org/atf/cf/%7BDEB6F227-659B-4EC8-8F84-8DF23CA704F5%7D/EEEethanol7.pdf.
Schindall 2007	Joel Schindall. Telephone conversation with Arjun Makhijani, January 12, 2007. Notes by Arjun Makhijani.
Schrag 2007	Daniel P. Schrag. "Preparing to Capture Carbon." *Science,* v.315 (9 February 2007), pages 812-813.
Schwarzenegger 2007	Arnold Schwarzenegger. *Transcript of Governor Arnold Schwarzenegger's State of the State Address as Delivered...Tuesday, January 9, 2007.* Press Release. [Sacramento, CA]: Office of the Governor, January 9 2007. On the Web at http://gov.ca.gov/index.php?/text/press-release/5089/.
Shepard and van der Linden 2001	Sam Shepard and Septimus van der Linden. "Compressed Air Energy Storage: Compressed Air Energy Storage Adapts Proven Technology to Address Market Opportunities." *Power Engineering*, v.105, no.4 (April 2001), pages 34-37. Web version at http://pepei.pennnet.com/Articles/Article_Display.cfm?Section=Archives&Subsection=Display&ARTICLE_ID=98019&KEYWORD=%22Norton%20Energy%20Storage%22.
Siemers 2007	Erik Siemers. "Expert Has Doubts About New Mexico Tire Recycler. *Albuquerque Tribune*, July 12, 2007. On the Web at http://www.abqtrib.com/news/2007/jul/12/expert-has-doubts-about-new-mexico-tire-recycler.

Smith 2006   Brice Smith. *Insurmountable Risks: The Dangers of Using Nuclear Power to Combat Global Climate Change.* Takoma Park, MD: IEER Press; Muskegon, MI: RDR Books, 2006. Summary on the Web at http://www.ieer.org/reports/insurmountablerisks/summary.pdf.

Smith and   Brice Smith and Arjun Makhijani. "Nuclear is Not the Way." *Wilson Quarterly,* v.30
Makhijani   (Autumn 2006), pages 64-68. On the Web at http://www.wilsoncenter.org/index.
2006   cfm?fuseaction=wq.essay&essay_id=204360.

Solion 2003   Solion Limited. *Li-ion Battery System for Solar Racing Cars.* London: Solion, [2003]. On the Web at http://www.lsbu.ac.uk/solarcar/solion.pdf.

Statistical   United States. Census Bureau. *The 2007 Statistical Abstract: The National Data Book.*
Abstract   [Washington, DC]: Bureau, 2007. On the Web at http://www.census.gov/compendia/
Online 2007   statab/.

              Table 767. Industrial Production Indexes by Industry and Major Market Groups: 1972 to 2005, at http://www.census.gov/compendia/statab/tables/07s0767.xls, with links from http://www.census.gov/compendia/statab/business_enterprise/economic_indicators/.

              Table 828. Cropland Used for Crops and Acreages of Crops Harvested: 1970 to 2005, at http://www.census.gov/compendia/statab/tables/07s0828.xls, with links from http://www.census.gov/compendia/statab/agriculture/.

              Table 897. Energy Consumption, by End-Use Sector: 1949 to 2005, at http://www.census.gov/compendia/statab/tables/07s0897.xls, with links from http://www.census.gov/compendia/statab/energy_utilities/.

              Table 1081. Domestic Motor Fuel Consumption by Type of Vehicle: 1970 to 2004, at http://www.census.gov/compendia/statab/tables/07s1081.xls, with links from http://www.census.gov/compendia/statab/transportation/.

Stern   Nicholas Stern. *Stern Review: The Economics of Climate Change: Executive Sum-*
Review 2006   *mary.* [London]: HM Treasury, October 30, 2006. On the Web at http://www.hm-treasury.
Executive   gov.uk/independent_reviews/stern_review_economics_climate_change/sternreview_
Summary   summary.cfm.

Stern Review   Nicholas Stern. *The Economics of Climate Change: The Stern Review.* Cambridge:
2007   Cambridge University Press, 2007. Links to the full report, first issued October 30, 2006, on the Web at http://www.hm-treasury.gov.uk/independent_reviews/stern_review _economics_climate_change/sternreview_index.cfm. The 2006 e-version has different page numbers than the printed book. We cite the book.

Suits 1951   C.G. Suits. "Power from the Atom – An Appraisal." *Nucleonics,* v.8, no.2, (February 1951), pages 3-8.

Sunlight   Sunlight Direct. *Sunlight Direct's HSL3000 for Commercial Lighting.* Knoxville, TN:
Direct 2005   Sunlight Direct, 2005. On the Web at http://www.sunlight-direct.com/HSL3000_Brochure.
              pdf.

Supreme   Supreme Court of the United States. *Syllabus: Massachusetts et al.* v. *Environmental*
Court 2007   *Protection Agency et al. Certiorari to the United States Court of Appeals for the District Of Columbia Circuit.* No. 05–1120. Argued November 29, 2006—Decided April 2, 2007. On the Web at http://www.supremecourtus.gov/opinions/06pdf/05-1120.
              pdf.

Tampa Elec-   Tampa Electric Company. *Biomass Test Burn Report Polk Power Station Unit 1.*
tric 2002   [Tampa, FL: TECO], April 2002. On the Web at http://www.treepower.org/TECO/polk-
              cofiring-testburn.pdf.

TEDB 2006   Stacy C. Davis and Susan W. Diegel. *Transportation Energy Data Book.* Ed. 25.
ORNL-6974 (Edition 25 of ORNL-5198). Oak Ridge, TN: Center for Transportation Analysis,
Engineering Science & Technology Division, Oak Ridge National Laboratory, 2006. Links to
the main report are on the Web at http://cta.ornl.gov/data/download25.shtml.

Chapter 2 Energy, at http://cta.ornl.gov/data/tedb25/Edition25_Chapter02.pdf.

Tesla Motors   Tesla Motors Web Site [San Carlos, CA]: Tesla, 2007. On the Web at http://www.
teslamotors.com/. Viewed March 1, 2007.

Electric Power, at http://www.teslamotors.com/performance/electric_power.php.

Tesla Motors Surpasses 500th 2008 Roadster Reservation (June 23, 2007), at http://
www.teslamotors.com/media/press_room.php?id=573.

*Tesla Roadster 'Signature One Hundred' Series Sells Out: First 100 Cars Quickly
Go to Aficionados Desiring Clean, Fast and Stylish Sports Car* (August 15, 2006),
at http://www.teslamotors.com/media/press_room.php?id=29.

Tesla Motors   Tesla Motors. *Performance Specs: Tesla Roadster Specifications.* [San Carlos, CA]:
2007   Tesla, 2007. On the Web at http://www.teslamotors.com/performance/specs.php.

Tilman, Hill,   David Tilman, Jason Hill, and Clarence Lehman. "Carbon-Negative Biofuels from Low-Input
and Lehman   High-Diversity Grassland Biomass." *Science*, v.314 (8 December 2006), pages 1598-1600.
2006

Tupolev 2006   Tupolev. *Cryogenic Aircraft: Development of Cryogenic Fuel Aircraft.* [Moscow]: PSC
«Tupolev», [2006]. On the Web at http://www.tupolev.ru/English/Show.asp?SectionID=82.

UN 2006   United Nations. *World Population Prospects: The 2006 Revision.* [New York]: UN Secretariat, Department of Economic and Social Affairs, Population Division, as of July 09, 2007.
On the Web at http://esa.un.org/unpp/.

UNFCCC   *United Nations Framework Convention on Climate Change.* [New York: UN, 1992. On the
1992   Web at http://unfccc.int/resource/docs/convkp/conveng.pdf. Viewed on July 9, 2007.

University   University of Chicago. *The Economic Future of Nuclear Power.* A Study Conducted at The
of Chicago   University of Chicago. [Chicago: University], August 2004. On the Web at http://www.anl.
2004   gov/Special_Reports/NuclEconAug04.pdf.

University   University of Delaware. *V2G: Vehicle to Grid Power.* [Newark]: University, [2006]. On the
of Delaware   Web at http://www.udel.edu/V2G.
V2G

USCAP 2007   United States Climate Action Partnership. *A Call for Action: Consensus Principles and
Recommendations from the U.S. Climate Action Partnership: A Business and NGO Partnership.* [Washington, DC]: USCAP, January 22, 2007. On the Web at http://www.us-cap.
org/ClimateReport.pdf.

US Census   United States. Census Bureau. *U.S. Interim Projections by Age, Sex, Race, and Hispanic
2004   Origin.* Table 1a. Projected Population of the United States, by Race and Hispanic Origin:
2000 to 2050. [Washington, DC]: The Bureau, March 18, 2004. On the Web at http://
www.census.gov/ipc/www/usinterimproj/natprojtab01a.pdf.

USGCRP   US Global Change Research Program. *US National Assessment of the Potential Conse-
2003   quences of Climate Variability and Change Educational Resources Regional Paper: Alaska:
Forests.* Washington, DC: USGCRP, updated 12 October, 2003. On the Web at http://www.
usgcrp.gov/usgcrp/nacc/education/alaska/ak-edu-5.htm.

USGS 2006	David A. Buckingham. *Steel Stocks in Use in Automobiles in the United States.* Fact Sheet 2005 – 3144. Denver, CO: U.S. Department of the Interior, U.S. Geological Survey, January 2006. On the Web at http://pubs.usgs.gov/fs/2005/3144/fs2005_3144.pdf.
Utah Geological Survey	Utah Geological Survey. *CO2 Sequestration Project Overview: Reactive, Multi-phase Behavior of CO2 in Saline AquifersBbeneath the Colorado Plateau.* [Salt Lake City]: The Survey, [no date]. On the Web at http://geology.utah.gov/emp/co2sequest/overview.htm.
Vedantam 2006	Shankar Vedantam. "Science Confirms: You Really Can't Buy Happiness," *Washington Post,* July 3, 2006; page A02. On the Web at http://www.washingtonpost.com/wp-dyn/content/article/2006/07/02/AR2006070200733.html.
Vienna Convention	*The Vienna Convention for the Protection of the Ozone Layer.* Nairobi: Ozone Secretariat, United Nations Environment Programme, 2001. On the Web at http://ozone.unep.org/pdfs/viennaconvention2002.pdf.
Walter et al. 2006	K.M. Walter, S.A. Zimov, J.P. Chanton, D.Verbyla, and F.S. Chapin. "Methane Bubbling from Siberian Thaw Lakes as a Positive Feedback to Climate Warming." *Nature,* v.443 (7 September 2006), pages 71 to 75.
Warrick 1999	Joby Warrick. "Evidence Mounts in Paducah." *Washington Post,* August 22, 1999, page A1. On the Web at http://www.washingtonpost.com/wp-srv/national/daily/aug99/paducah22.htm.
Weisman 2004	Steven R. Weisman. "U.S. in Talks With Europeans on a Nuclear Deal With Iran." *New York Times,* October 12, 2004.
WGA 2006	Western Governors' Association. Clean and Diversified Energy Initiative. *Solar Task Force Report.* [Denver]: WGA, January 2006. On the Web at http://www.westgov.org/wga/initiatives/cdeac/Solar-full.pdf.
Wigley and Raper 2001	T.M.L. Wigley and S.C.B. Raper. "Interpretation of High Projections for Global-Mean Warming." *Science,* v.293 (20 July 2001), pages 451-454.
Wikipedia Hindenburg 2007	Wikipedia. *LZ 129 Hindenburg.* Last modified 9 March 2007. On the Web at http://en.wikipedia.org/wiki/Hindenburg_disaster.
Wilson, Johnson, and Keith 2003	Elizabeth J. Wilson, Timothy L. Johnson, and David W. Keith. "Regulating the Ultimate Sink: Managing the Risks of Geologic CO_2 Storage." *Environmental Science and Technology,* v.37, no.16 (2003), pages 3476-3483. On the Web at http://www.ucalgary.ca/~keith/papers/54.Wilson.2003.RegulatingTheUltimateSink.e.pdf.
Winkler 2006	Michael Winkler. *Benefits for Renewable Energy of Integrating the Transportation and Heating Sectors with the Electric Grid.* June 1, 2006. Unpublished. Winkler is affiliated with Schatz Energy Research Center, Humboldt State University.
Winkler 2007	Michael Winkler. *Zero-Net Energy Home (We Generate As Much Energy As We Use).* Arcata, CA, Carol McNeill and Michael Winkler, 2007. Winkler is affiliated with Schatz Energy Research Center, Humboldt State University.
Wolverton and McDonald 1977	B.C. Wolverton, Rebecca C. McDonald. "Wastewater Treatment Utilizing Water hyacinths (*Eichhornia crassipes*) (Mart) Solms." NASA-TM-87544. In *Treatment and Disposal of Industrial Wastewaters Residues: Proceedings of the 1977 National Conference on Treatment and Disposal of Industrial Wastewaters and Residues, April 26-28, 1977, Houston, Texas,* pages 205 to 208. On the Web at http://ntrs.nasa.gov/archive/nasa/ssctrs.ssc.nasa.gov/tm-87544/tm-87544.pdf.

Wolverton and McDonald 1979	B.C. Wolverton and Rebecca C. McDonald. *Energy from Aquatic Plant Wastewater Treatment Systems.* NASA Technical Memorandum TM-X-72733. NSTL Station, MS, Earth Resources Laboratory, National Aeronautics and Space Administration, National Space Technology Laboratories, September 1979. On the Web at http://ntrs.nasa.gov/archive/nasa/ssctrs.ssc.nasa.gov/tm-x-72733/tm-x-72733.pdf.
Wu et al. 2001	X. Wu, R.G. Dhere, D.S. Albin, T.A. Gessert, C. DeHart, J.C. Keane, A. Duda, T.J. Coutts, S. Asher, D.H. Levi, H.R. Moutinho, Y. Yan, T. Moriarty, S. Johnston, K. Emery, and P. Sheldon. *High-Efficiency CTO/ZTO/CdS/CdTe Polycrystalline Thin-Film Solar Cells.* Preprint. Conference Paper. NREL/CP-520-31025. To be presented at the NCPV Program Review Meeting, Lakewood, Colorado, 14-17 October 2001. Golden, CO: National Renewable Energy Laboratory, October 2001. On the Web at http://www.nrel.gov/docs/fy02osti/31025.pdf.
Yergin 1991	Daniel Yergin. *The Prize: The Epic Quest for Oil, Money & Power.* New York: Simon & Schuster, 1991.
York 2001	Herbert York. Interview with Arjun Makhijani, 23 June 2001. Interview notes reviewed and corrected by Dr. York.
Zagórze, no date	RUSE. *Zagórze.* [Besançon, France]: Energie-Cités, RUSE, [n.d.]. On the Web at http://www.ruse-europe.org/IMG/pdf/Zagorze_EN.pdf.
Zirschky and Reed 1988	John Zirschky and Sherwood C. Reed. "The Use of Duckweed for Wastewater Treatment." *Journal – Water Pollution Control Federation,* v.60 (July 1988), pages 1253-1258.
Zerriffi and Annie Makhijani 2000	Hisham Zerriffi and Annie Makhijani. *The Nuclear Alchemy Gamble: An Assessment of Transmutation as a Nuclear Waste Management Strategy.* Takoma Park, MD: Institute for Energy and Environmental Research, August 25, 2000. On the Web at http://www.ieer.org/reports/transm/report.pdf. (May 12, 2005 Web-posted version with modifications to Appendix B).
Zwetzig 2007	Randy Zwetzig. Telephone interview with Arjun Makhijani, August 22, 2007. Notes of Arjun Makhijani.

INDEX

Africa 2, 19, 47, 204

AFS Trinity 68, 214, *See also* hybrid, extreme

Airbus 86, 87, 88, 215,

Air conditioning 73, 93, 106, 117, 216

Aircraft: efficiency 88, 216
fuel for 84-88,
hydrogen fuel for 54, 86-88, 128, 156, 215

Air pollution xv, 26, 99, 112, 114, 120, 166
costs 26

Alabama 70

Algae 21, 49, 51, 52, 59, 60, 85, 86, 179, 198, 199, 200. *See also* microalgae

Allowances, revenues from the sale of 142

Altairnano 64, 66, 169, 214,

Amazonian rainforest xvi

Antarctic Ozone Hole 125

Appalachian region 143, 174

Appliances, efficiency standards 6, 12, 130, 131, 149, 175

Aquatic plants *See* biomass, aquatic

Arctic ice xvi

Arizona Public Service 21

Atomic Energy Commission xviii, 171, 182, 184,

Atoms for Peace xix, 188

Auctioning *See* emissions allowances

Automobiles:
all-electric 12, 17, 23, 42, 63-67, 80, 98-100, 104, 152, 211, 214, 215
efficiency standards 12, 23, 89, 98, 130, 131, 139, 149, 156, 175, 195, 202
hybrid 23, 42, 63, 67, 68, 98, 143, 214
installed power 23, 64
plug-in hybrid 23, 26, 42, 63, 66, 68, 152, 155, 212
V2G 23, 64-66, 180

Bacher 182, 218

Backfitting 81

Backup technologies *See* Redundancy

Bahrain xviii, 170

Bangladesh xvi

Batteries:
capacity 23
charging and discharging 23, 64, 66, 67, 68, 169, 211
cycles 23, 66, 169
depreciation 23, 66
lead-acid 63, 64, 68, 117
lithium-ion 17, 23, 45, 63, 66, 67, 68, 83, 99, 112, 117, 118, 127, 128, 153, 155, 156, 169, 211, 214, 215, 218
NiMH 63
sodium-sulfur 68

Bay St. Louis, Mississippi 55

Berzin, Isaac xiv, 21, 26, 50, 210, 213, 214

Bio-methane 104

Biobutanol 26, 198

Biodiesel 84, 85

Biofuels:
from food crops 48, 58, 175
from corn 45-48, 52, 172, 199, 212
from palm oil 48, 153, 172
gaseous 96,107-112, 117, 118, 127, 147, 148, 174
hydrogen to displace 113
jet aircraft 84, 85, 100, 152
and requirements 53, 111, 112, 127, 148, 151, 154, 174, 216
liquid 52, 60, 84, 107-112, 117, 118, 126-128, 147, 148, 174
See also ethanol, biodiesel, biomethane,biogas

Biomass:
-based biofuels 30
and coal 54
aquatic 49, 59, 110, 154
cellulosic 58
high productivity 49, 54, 55, 56, 109, 110, 111, 128, 150, 172
liquid fuels from 48
solid 53, 58, 60, 104, 105, 108, 109, 148

Bioreactors 51, 52

Bird kills 36

Blackout, 1965 103

BMW 113

Braking, regenerative 12, 68

Brazil 2, 45, 170, 205

Coal x, 5, 8, 10, 15, 18-21, 24, 28, 47, 49, 51-54, 59, 69,-72, 82, 101, 111, 114, 119, 121-123, 130, 131, 136, 143-146, 149, 151, 155, 159, 169, 173, 175, 179, 182, 190, 191, 197, 204, 208-210, 214, 217
  -fired power plants 19, 21, 51, 101, 130, 131, 144, 146, 149, 169, 175, 191, 210, 214
    ban 130, 131, 149, 175
    co-firing biomass with 53, 54
    gasification 19, 169, 173
    pulverized 18- 20, 24, 144, 173

Coefficient of performance 116, 215

Cole, Sterling 182

Colorado 30, 31, 34, 72

Combined cycle natural gas plant vi, 19, 59, 118, 150, 173, 179

Combined heat and power *See* CHP

Commercial Energy vi, 80, 96

Commonwealth Edison 183

Compact fluorescent lamps 12, 83, 129, 178

Contingency plan 118, 128

COP *See* Coefficient of performance

Coral reefs xvi

CORINAIR 133

Corn:
  flour 47
  from ethanol 45, 52, 172
  Grupo Maseca 47
  Minsa 47
  solar energy capture 46, 49
  stover 48

Costs
  distribution 117
  storage 40, 44, 216
  transformation of the vehicular sector 162
  transmission 124

Cow dung 45

Crop residues 45, 48, 49, 58, 154, 213

Crude oil. *See* oil

CRYOPLANE 86, 87

Declaration of Independence xxi

Defense, Department of 85. *See also* Pentagon, military spending

Defense Advanced Research Projects Agency (DARPA) 85

Delaware, University of 64, 214

Delivered energy 17, 18, 73-75, 77, 78, 80, 95, 96, 97, 100, 102, 103, 107, 109, 115, 136, 139, 167, 171
  per square foot 139

Demonstration plants 49, 139, 149, 150, 151, 153, 154, 155

Denmark 34

Densely populated communities 92

Detroit Edison 183

Developing countries 2, 3, 45, 48, 125, 130, 134, 135, 164, 167, 172, 205, 212, 213

Distillation 94, 157, 197

Docking station 23

DOD *See* Defense, Department of

DOE 39, 60, 96, 113, 114, 115, 186, 189, 210-213, 216, 218

Duckweed 49, 56, 59, 110, 150, 172

Duke Energy 20

DuPont xiii, 93, 94, 101, 125, 136, 193, 195, 196, 198-201, 206, 209- 211

Durant Road Middle School 79

Efficiency of transit bus 90, 91

Efficiency, manufacturing 94, 197

Efficiency standards:
  appliances 6, 12, 130, 131, 139, 149,175
  buildings 82, 130, 131, 139, 149, 175
  vehicles 23, 89, 98, 130, 131, 139, 149, 167

Electricity: transmission and distribution losses 102, 107

Electricity generation, thermal losses in 79

Electricity prices:
  peak 37
  retail 7
  time-of-use 21, 40, 124, 153

Electric Power Research Institute 61, 133

Electrolysis 112-114, 148, 159, 200

Electrolyzer: cost 114
  efficiency 114

Emission *See* CO_2 emissions
  per-capita 3, 9, 10
  tailpipe stands 167

Emissions allowances 18, 92, 101, 135, 138, 152, 167, 173, 210
  free 131, 134, 135

Endangered species xvi, 63

Geologic disposal, deep 20

Geothermal energy: conventional hydrothermal 61
  hot rock 30, 62, 152, 154

Geothermal heat pump *See* Heat pumps

Germany xxii, 1, 34, 38, 70, 132, 208

Glaciers xvi

Global Nuclear Energy Partnership xviii, 88, 189
  and NPT 188

Global positioning systems 65

Global warming 2, 15, 71, 131, 135, 146, 205, 208, 217

Glucose 59

GNEP *See* Global Nuclear Energy Partnership

Google 26, 41, 42, 66, 67, 211, 212, 214

Gore, Al xxii, 2, 206, 209, 219

Grasses *See* switchgrass and prairie grasses

GreenFuel xiv, 26, 50, 210

Greenhouse gas emissions xv, xvii, xxii, 1, 2, 12, 46, 47, 48, 58, 72, 80, 87, 93, 94, 101, 146, 172, 190, 206, 207, 209, 210, 212

Green Wind Farm 30

Grid:
  contingencies for the electric vi, 118
  distributed 28, 103, 141, 147, 178
  smart 107, 141, 150, 159, 169

Gross Domestic Product. *See* GDP

Gross State Product 9, 10, 11

Ground-source heat pumps. *See* heat pumps

Grupo Maseca 47

Gulf Cooperation Council (GCC) xviii, 170

Halocarbon emissions 94

Hanover House 76, 77, 97, 115, 116

Hard cap. *See* cap

Hawaii 46, 61

Health x, xvii, xxi, 55, 88, 92, 103, 114, 120, 128, 166, 171, 188, 189
  air pollution 99, 114, 166
  spent fuel xvii, 171, 189

Heating and cooling requirements 10

Heating Loads 76

Heat pump:
  earth-source 77, 80, 178

efficiency 77, 80, 178
  geothermal 77, 178
  ground-source 77, 178

Heat storage: capital costs 44

Heavy metals 44, 55-57
  pollution 55

High-level waste. *See* nuclear waste

Housing:
  low-cost 81
  new 151
  backfits 81, 82

Hurricane Katrina 164

Hutchins, Robert 182

HVAC 78, 197, 215

Hybrid, extreme 68

Hybrid, plug-in 23, 26, 42, 66, 68, 152, 155, 212

Hydrocarbons 60, 87, 112, 148, 199

Hydrocarbons, unburned 60, 87, 112

Hydrogen production:
  biological 59
  centralized 112, 115, 149
  direct 30, 148
  direct from solar energy 30, 52, 53, 59, 60, 61, 127, 128
  distributed 113-115, 127, 143, 174
  electrolytic 127, 143, 155,175, 178
  high-temperature 60, 173
  infrastructure 112, 115, 127, 149
  photoelectrochemical 60, 115, 173, 175
  photolytic 59, 60, 154, 175, 178
  solar-energy-driven 59, 173
  wind 112-115, 151

Hydropower 63, 101, 102, 104, 106, 122, 150, 153, 168, 213

IEER *See* Institute for Energy and Environmental Research

IGCC *See* Integrated Coal-Gasification Combined Cycle

Inconvenient Truth xxii

Independent System Operator 65

India 2, 19, 185, 188, 191, 205, 206
  1974 nuclear explosion 185
  zero CO_2 emissions goal 205

Indonesia 48, 172. *See also* palm oil

Insolation 43, 44, 46, 111, 168
  diurnal variations 43, 44
  seasonal variations 43, 44, 46

Institute for Energy and Environmental Research i, ii, xiv, xvii, 16, 62, 79, 80, 98, 105, 111, 122, 161, 162, 163, 208, 217, 218

Insulation 75, 77, 78, 80, 81, 94, 125, 140, 196, 197

Insurance:
nuclear power xviii, 20, 144, 145, 149, 171, 175
subsidies 20
subsidized housing 151

Insurance companies 146, 206
and climate change 206

Integrated Coal-Gasification Combined Cycle 19, 20, 24, 53, 54, 56, 60, 72, 104, 145, 151, 152, 154, 159, 169, 173, 179, 210

Intergovernmental Panel on Climate Change xvi, xvii, 1, 2, 166, 203

Intermittency 21, 31, 32, 104, 145
of solar 43
of wind 32, 61, 64
of wind and solar 21, 30, 115, 168

Internal combustion vehicles 60, 100, 113, 114, 126, 148, 151, 174

International law 3

Iowa 31, 70, 151

IPCC See Intergovernmental Panel on Climate Change

Iran xviii, 170, 204

Iraq xx, 25

Iraq Study Group xx

Island countries xvi

Israel xviii, xix, 170

Japan 1, 3, 4, 45, 164, 170, 204
nuclear weapon debate 188
per person emissions 2
See also Tokyo Electric

Japan Focus vii, xiv, 203, 219

Jet fuel See JP-8 and aircraft fuel

JP-8 and aircraft fuel 84-88
surrogate 85, 86

Kansas 10, 31, 168

Kerosene 84-87, 126. See JP-8

Kissinger, Henry xx, 25, 208

Korea 170

Kuwait xviii, 170

Kyocera 41

Lamar, Colorado 30

Lamps, fluorescent, ballasts for 79

Land-area requirements ˉ108, 109-111, 114, 154, 216

Landfill gas 59, 109, 217

Leadership in Energy and Environmental Design. See LEED

Lead times xvii, 189

LEED 151, 160, 162, 179

Lifestyles v, xx

Light emitting diodes 83

Lighting 74-76, 78, 80, 81, 83, 93, 96, 117, 129, 157, 173, 194, 215
efficiency 76-79, 82
hybrid 83, 215

Liquefied natural gas 22, 84, 86

LNG See liquefied natural gas

Load:
following 33, 35
off-peak 21, 22, 24, 26, 42, 52, 64, 66, 68, 69, 70, 99, 106, 113, 155, 163, 173, 179, 213
on-peak 70, 106, 163
peak 70, 81, 106
profile 81, 82

London 88, 120, 125, 206

Los Alamos National Laboratory 61

Los Angeles i, xi, xiii, 88, 91, 92, 215

Losses
biofuel 111, 112, 147, 158, 171
biofuel production 107
distribution 102
transmission 102

Louisiana 43, 49, 50

Luz International 44, 62

Manhattan, car ownership in 89

Manhattan Project 182, 183

Marlboro, Massachusetts 38

Massachusetts Institute of Technology xiv, 20, 21, 49, 50, 51, 198, 213

McIntosh, Alabama 70

Mercury 82

Merkel, Angela 1

Metals production 92